W9-ADO-595

Aviation

A Smithsonian Guide

6
U.S.NAVY

Blue Angels

A Smithsonian Guide

Aviation

A Smithsonian Guide

Donald S. Lopez
Senior Advisor to the
Director of the National Air
and Space Museum

Macmillan • USA

Macmillan • USA
A Prentice Hall Macmillan Company
15 Columbus Circle
New York, NY 10023

A Ligature Book

All the information in this book is correct according to the best current resources.
Any questions or comments are welcome and should be directed to: Ligature, Inc.,
Attn: Trade Department, 75 Arlington St., Boston, MA 02116

In the Gallery

pages 2–3. The sky was filled with a wide variety of flying machines at this 1910 air show, although the photo was retouched to bring all of them close together.
pages 4–5. The U.S. Air Force's Thunderbirds maneuver at high speed in tight formation.
pages 6–7. Two of the U.S. Navy's Blue Angels become mirror images as one flies inverted.
page 8. The powerful Gee Bee R-2 Super Sportster makes a high-speed turn.
page 11. The sleek F-15 Eagle was the premier long-range fighter in the Persian Gulf War.
page 12. The U.S. Air Medal is awarded for military service.

Library of Congress Cataloging-in-Publication Data

Lopez, Donald S., 1923–
Aviation: A Smithsonian Guide / Donald S. Lopez.
p. cm.
"A Ligature book."
Includes bibliographical references and index.
ISBN 0-02-860006-1. ISBN 0-02-860041-X (pbk.).
1. Aeronautics—History. 2. Aeronautics, Military—History
I. Title.
TL515.L67 1995
629.13'009—dc20 94-32992 CIP

Ligature Inc.

Publisher
Jonathan P. Latimer

Series Design
Patricia A. Eynon

Editorial
Susanna Brougham
Mary Ashford
Elizabeth A. Mitchell
Steven Thomas

Design
Melissa Niederhelman

Research
Michael Pistrich
Kristen Holmstrand

Production
Anne E. Spencer
Paul Farwell
Ron Frank

Contributors
Holly Manry
Tammy Zambo

The Smithsonian Institution

Patricia Graboske
Melissa Keiser
Timothy J. Cronen
Jim Wilson
Kristine Kaske
Karen Whitehair

Macmillan

Robin Besofsky

Macmillan books are available at special discounts for bulk purchases, for sales promotions, fund-raising, or educational use. For details, contact:

Special Sales Director
Macmillan Publishing Company
15 Columbus Circle
New York, NY 10023

Manufactured in Hong Kong
10 9 8 7 6 5 4 3 2 1

EG
095
EG
58
EG
097

Contents

Dreams, Failures, and Triumphs 14–39

Winged Gods and Dreams of Flight • Balloons and Gliders • The Wright Brothers and Powered Flight • Skeptics and Pioneers • International Competitors

Featured Plane: Wright 1903 Flyer

World War I 40–61

The War in the Air • The Fokker Scourge • Dogfights and Air Tactics • Von Richthofen, Rickenbacker, Fonck, and Bishop • Bombers and a New Kind of War

Featured Planes: SPAD XIII and Albatros D.Va

The Golden Age 62–87

Stunt Fliers and Exhibition Teams • Barnstorming, Air Races, and New Records • Lindbergh, Earhart, and Post • Flying Around the World

Featured Plane: *Spirit of St. Louis*

Air Transport 88–105

Airmail and Passenger Services • The First Commercial Airlines • Crop Dusting and Aerial Advertising • The Crash of the *Hindenburg*

Featured Plane: Douglas DC-3

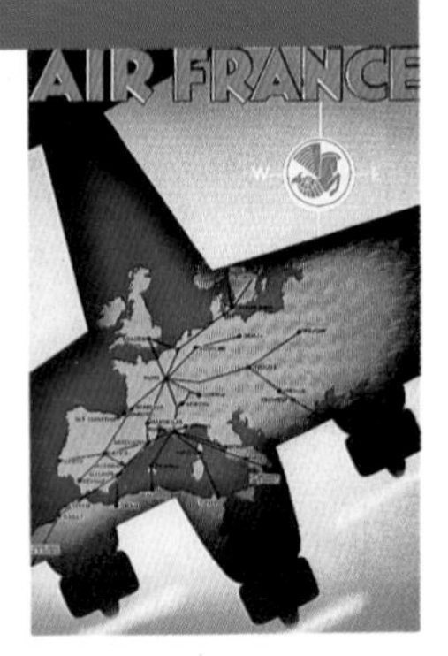

World War II 106–139

Blitzkrieg and the Luftwaffe • The Battle of Britain • Pearl Harbor and War in the Pacific • Bombing Raids in Europe • Kamikazes and Atomic Bombs

Featured Planes: Messerschmitt Me-109 and P-51 Mustang

The Jet Age 140–177

Jet Fighters in World War II • Breaking the Sound Barrier • The Korean War • Supersonic Fighters • Jet Bombers and Transports • Jet Airliners • Vertical Flight

Featured Plane: Concorde

Wild Ideas and Future Flight 178–208

Experiments and Flops • Homebuilts and Ultralights • The Skunk Works • Vertical Takeoff and Landing • New Wings, New Propulsion, New Controls

Featured Planes: Stealth Fighter and Stealth Bomber

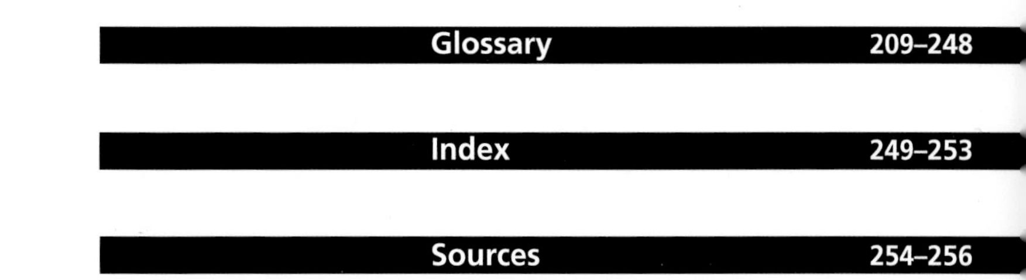

Glossary 209–248

Index 249–253

Sources 254–256

Dreams, Failures, and Triumphs

Flight has long tantalized the human imagination, and this fascination inspired some outlandish inventions: flapping devices, bird-driven flying machines, and contraptions with sails and oars. During the nineteenth century, experiments with gliders and studies in aerodynamics led at last to the historic moment in 1903 when the Wright brothers achieved powered, sustained, piloted flight over Kitty Hawk.

This colorful replica of the Montgolfier balloon was flown in 1983 to celebrate the 200th anniversary of the first balloon flight, which took place in France in 1783.

In a well-known Greek myth, Daedalus builds wings of feathers and wax for himself and his son, Icarus, so that they can escape from prison by flying. Daedalus warns his son not to fly too high, lest the heat of the sun melt his wings. But in the joy of flight Icarus forgets the warning. He flies too high and plummets to his death.

Winged Gods and Dreams of Flight

Since ancient times, high places have been considered sacred—home to the gods, but beyond the reach of humans. Costumes and art from many cultures depict winged gods and humans who could ascend into the air. The dream of flying has long fired the human imagination.

Storytellers and experimenters contributed to the lore of flight. In Greek mythology, Perseus flew by using a pair of winged shoes lent by the god Mercury, and Bellerophon ascended into the heavens on the winged horse Pegasus. Flying carpets of Persian legend and demigods in India's myths were perfectly at home in the skies. But early efforts at human flight are tinged with the ridiculous rather than the divine. Lacking aids such as winged horses, people tried to imitate birds.

The history of human flight begins with adventurous people equipped with flapping wings who flung themselves from high places. These devices, called *ornithopters*, did not perform well. For example, in 1010 a monk named Eilmer jumped from Malmesbury Abbey in England and broke his legs. In 1162, a man in Constantinople fashioned sail-like wings from a fabric gathered into pleats and folds. He plummeted from the top of a tower and died. In 1536, Denis Bolori in France tried to fly using wings flapped by a spring mechanism. He fell to his death when the spring broke.

This detail from a painting by Piero di Cosimo, an Italian artist of the early Renaissance, depicts the Greek hero Perseus wearing his winged sandals and flying to the rescue of Andromeda.

"A bird is an instrument working according to mathematical law . . . which it is within the capacity of man to reproduce."
— Leonardo da Vinci

The Hindu god Vishnu rides on the back of Garuda, accompanied by his consort. The great bird Garuda agreed to become Vishnu's vehicle in return for the boon of eternal life. The pairs of birds symbolize the union of Vishnu and his wife.

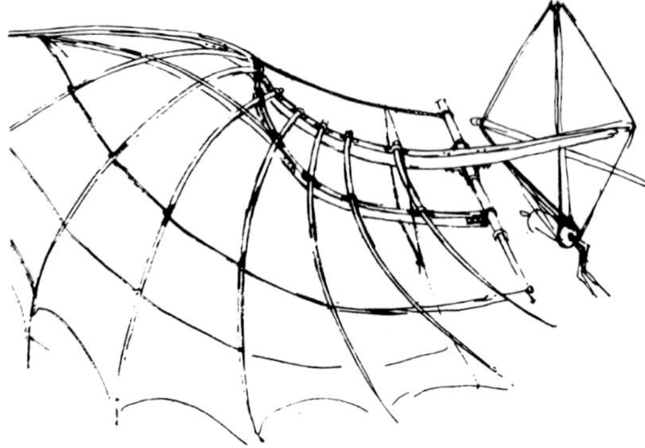

Leonardo da Vinci's notebooks, written between 1488 and 1514, include drawings and descriptions of many flying machines. This sketch reflects his study of bird wings and bat wings.

Leonardo da Vinci

In the early 1500s, Leonardo da Vinci had intuitively grasped a better understanding of flight than many experimenters did in later centuries. For example, he comprehended the importance of stability in a mechanism for flight. His notebooks contain sketches of a number of ornithopters and even a rudimentary helicopter. In one of his designs, the pilot lies on a board and manipulates cords and rods to move the wings. It is not known whether Leonardo tested his ideas, but he had advice for anyone who did: "You will experiment with this instrument on a lake . . . so that in falling you will come to no harm."

lloons and Lighter-an-Air Flight

Joseph and Etienne Montgolfier invented the hot-air balloon in France, in 1783. They built a globe-shaped balloon of taffeta and filled it with smoke from a fire made from moist straw and chopped wool. The hot air within the balloon, which was lighter than the surrounding air, lifted the balloon from the ground.

In September 1783 the Montgolfiers held a demonstration for the king and queen of France, Louis XVI and Marie Antoinette, at Versailles. The first passengers were a sheep, a rooster, and a duck. The strange upper atmosphere was blamed when the rooster sustained an injury in the air. (Actually, the sheep was responsible.)

In November 1783 the first human flight in history also took place at Versailles. Two men, J. F. Pilâtre de Rozier and the Marquis d'Arlandes, ascended in a Montgolfier balloon and flew over Paris, dousing burning holes in the balloon with sponges. The flight was witnessed by Benjamin Franklin and awestruck Parisians, who prayed for the safety of the airborne men.

Also in 1783, J. A. C. Charles made a balloon in France out of rubberized silk and filled it with hydrogen, freeing balloonists from the danger of an open fire. The first human flight in a hydrogen balloon occurred on December 1, 1783. In 1785, a piloted balloon was flown across the English Channel.

Gliders and Heavier-Than-Air Flight

An ornithopter's beating wings guaranteed a swift descent to earth. Fortunately, inventors in the 1800s turned their attention to studying the shape of the wing and how it performed in a stream of air. The result was the glider.

George Cayley

George Cayley, England's "father of aeronautics," was gifted with boundless intellectual curiosity. His studies ranged from the design of lifeboats to engines powered with gunpowder. He clarified principles that eventually led to heavier-than-air flight.

Perhaps Cayley's most profound insight was separating the concepts of lift, propulsion, and control. Though a bird's wings effortlessly provide all three functions, human flight could best be approached by studying and implementing each separately. Cayley is credited with the first major breakthrough in heavier-than-air flight: understanding wing shape and lift. By 1809 Cayley had developed a number of fixed-wing model gliders stabilized by tail assemblies. Full-scale gliders soon followed.

In one of these gliders, in 1853, legend has it that Cayley's coachman became the first person to glide in the air successfully. Shaken by the short flight, he abruptly gave notice, telling Cayley he had been hired to drive and not to fly.

William S. Henson and John Stringfellow greeted Cayley's work more enthusiastically. Using Cayley's calculations, in 1847 they produced a model craft with a 20-foot span, powered by a steam engine—then they quickly announced plans for a transit company to carry goods to China. Though their ambitions overshot their accomplishments (their design never flew), Henson and Stringfellow helped popularize Cayley's vision of a fixed-wing airplane.

George Cayley's sketches show a variety of gliders with fixed, curved wings. The glider at the bottom, which Cayley called a "governable parachute," is probably the type that achieved a successful flight in 1853, piloted by Cayley's anxious coachman.

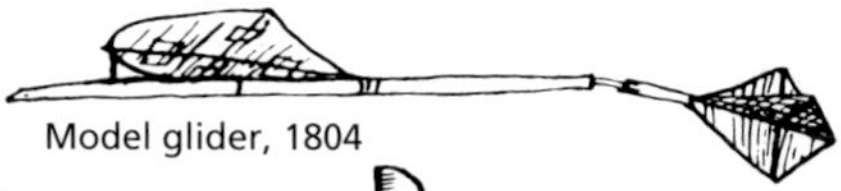

Model glider, 1804

Fixed-wing glider, 1849

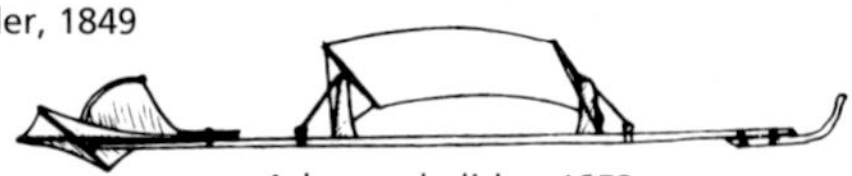

Advanced glider, 1853

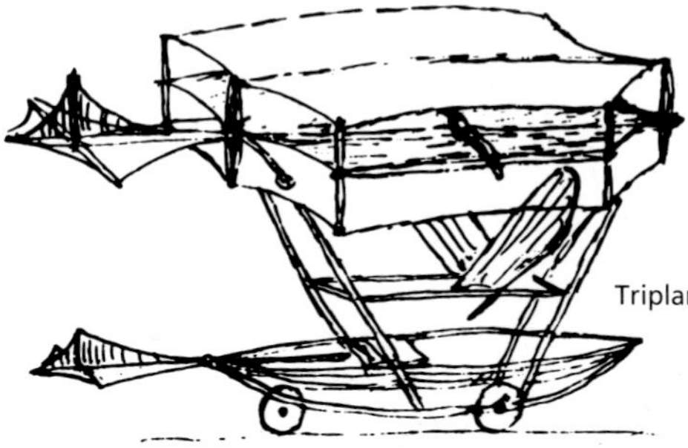

Triplane, 1849

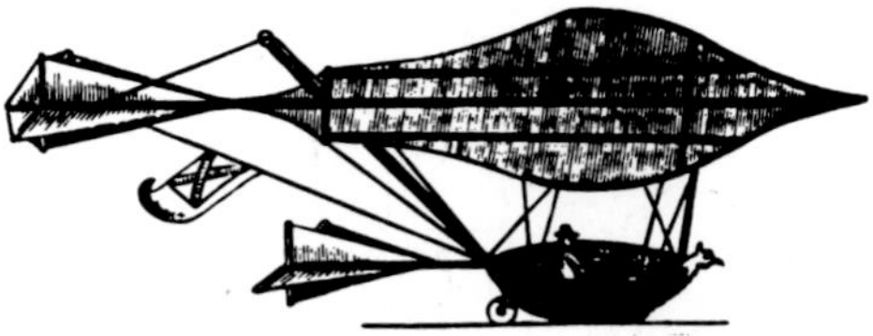

Man-carrying glider, 1852

Lilienthal takes to the air in a biplane glider, from his custom-built, cone-shaped hill located near Berlin. The cone shape allowed him to take advantage of the wind no matter which way it was blowing.

In 1891, after studying bird flight and aerodynamic theory for two decades, Otto Lilienthal began to test his ideas about wing design and lift by building and flying a series of gliders.

Otto Lilienthal

Between 1891 and 1896, Otto Lilienthal, a German civil engineer, completed nearly 2,000 flights with 18 glider designs. Both his single-wing and double-wing gliders resembled the wings of a soaring bird.

Lilienthal constructed his gliders from common materials, such as willow rods and waxed cotton. He launched them into the wind from hills, including a cone-shaped hill built especially for this purpose. To fly the glider, Lilienthal would crawl under the craft, position his arms in a set of cuffs, grasp a bar near the forward edge of the wings, and run down the slope. Once he was aloft, his legs dangled free.

An observer described Lilienthal's workmanship:

> So perfectly was the machine fitted together that it was impossible to find a single loose cord or brace, and the cloth was everywhere under such tension that the whole machine rang like a drum when rapped with the knuckles.

When Lilienthal took off, "the wind playing wild tunes on the tense cordage of the machine," the image was unforgettable: "the spectacle of a man supported on huge white wings, moving high above you at racehorse speed."

The fatal flaw of Lilienthal's gliders was lack of control: he balanced the craft by shifting his weight, reacting to its movements rather than directing them. If the glider went nose-up before the pilot could compensate, it fell out of control.

Despite this problem, Lilienthal flew his gliders with superb skill. Dubbed the first flying man, he became well known for his study and practice of the art of flying and his ability to glide for distances up to 1,000 feet.

On August 9, 1896, Lilienthal crashed in a glider from a height of over 50 feet and died soon after. His published studies and carefully documented test flights inspired many experimenters in flight.

Of the 18 models he tested, Lilienthal considered this glider, constructed in 1894, the most successful. Its willow and bamboo frame, covered by taut cotton fabric, measures 26 feet in width.

Octave Chanute

Octave Chanute, an American engineer, also studied heavier-than-air flight. Chanute's impact on aviation is all the more amazing because he spent so much time in other endeavors—constructing railroads, bridges, and stockyards, for example.

His experiments with Lilienthal-type gliders in 1894 and 1895 brought him to the conclusion that wing pivoting provided better control than shifting the pilot's weight. During 1896, he built five gliders, starting with a five-wing design and ending with two wings (a biplane). He theorized that a glider's wings should pivot on a central frame, rotating slightly if a gust of wind should strike the craft. Later he switched to rigid wings, held in position by trusses similar to those he used in bridge construction. Each 16-foot wing was covered with varnished silk. The pilot was suspended from two bars that ran down from the upper wings and passed under the pilot's armpits.

Chanute's success with gliders electrified the American aeronautical community. He supported and inspired the efforts of many experimenters in aviation, including the Wright brothers.

On October 7, 1903, the first launch of Samuel Langley's Aerodrome ended in the Potomac River. Langley's second attempt, on December 8, ended the same way.

THE WASHINGTON POST, FRI

LANGLEY EXPLAINS FIASCO

Buzzard's Collapse Wholly Due to

POLICE ARE BLAMED.

Verdict of Coroner's Jury Regarding the Death of John G. Carr.

This cartoon, published in the *Washington Post on* October 9, 1903, shows the derision Langley faced after the failure of the Aerodrome's flight attempts. Langley, in a top hat, perches on "The Buzzard"; pilot Charles Manly wears diving gear.

The Skeptics

Though long familiar with gliders and balloons, many Americans could not imagine flight by a powered, piloted aircraft. "Flight was generally looked upon as an impossibility," Orville Wright recalled.

Leading periodicals discussed flying machines in a condescending fashion. The *New York Telegraph* scoffed at the idea of sending mail by air: "Love letters will be carried in a rose-pink aeroplane steered with Cupid's wings and operated by perfumed gasoline." Newspapers were especially brutal to Samuel P. Langley after the failure of the Aerodrome.

Even in 1906, well after the Wrights' early successes, established scientists such as astronomer and mathematician Simon Newcomb could confidently proclaim that powered, piloted flight in a heavier-than-air craft was simply impossible.

Powered Flight

Samuel P. Langley and the Aerodrome

The third secretary of the Smithsonian Institution, Samuel P. Langley, tackled the question of adding power to workable models. In 1887, Langley began his studies by constructing a series of small flying models powered by twisted rubber bands. He experimented with many combinations of wings, fuselages, propellers, and tail assemblies until he was ready to move on to full-scale flying machines.

Langley called these machines "aerodromes" and launched them from a houseboat anchored in Washington D.C.'s Potomac River. After a number of unsuccessful efforts, in May 1896 two of Langley's unpiloted, powered aerodromes made history. Their flights (up to 4,200 feet in length) marked the first time that full-scale, heavier-than-air, powered models flew rather than hopped.

Because of this success, Langley was able to obtain substantial government funding to build a powered aircraft capable of being flown by a pilot. After years of study (and highly publicized spending), Langley prepared to conquer piloted flight. In October 1903 he catapulted the Aerodrome, piloted by Charles Manly and powered by a 52-horsepower gas engine, over the Potomac from his houseboat. The Aerodrome's wings collapsed abruptly—falling, as a reporter noted, "like a handful of mortar"—and dumped Manly into the water. At the next launch, in December, the pilot catapulted forward only to find himself looking straight up into the sky as the plane crashed into the water on its back. The pilot was not seriously injured, but Langley's involvement in aviation was over for good.

Samuel P. Langley, secretary of the Smithsonian Institution, poses with Charles Manly (in white). Manly, a talented engineer, helped build the Aerodrome's engine. He survived two plunges into the Potomac as the Aerodrome's pilot—his first attempts at flying.

Wing Design

The shape of a wing determines how it will perform in a stream of air. A flat shape fights air flow, causing drag (resistance). A curved shape allows air to flow smoothly around it.

A wing that is curved on top and flatter on the bottom creates lift. The air passing over the top surface has a longer distance to travel and therefore moves more quickly, creating less pressure than the slower air flowing below the wing. The air below the wing exerts pressure upward, causing lift.

If the wing is tilted upward (when the nose of the plane points up), lift increases. But too steep a tilt causes wing stall, or a complete loss of lift.

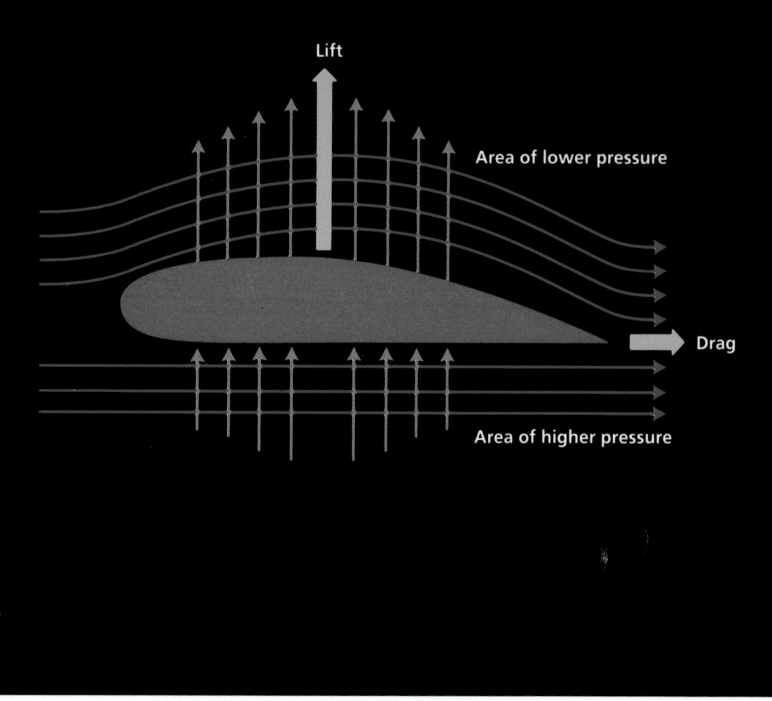

Pioneers Par Excellence: The Wright Brothers

"From the time we were little children my brother Orville and myself lived together, worked together and, in fact, thought together." In this way Wilbur Wright summed up one of the most famous collaborations in history. The Wrights' interest in aviation was kindled by a toy helicopter, which their father, Bishop Milton Wright, gave them as a gift in 1878.

The brothers never hesitated to pursue their diverse interests. As young adults they roamed from one endeavor to the next, challenging their natural mechanical aptitudes and restless intellects. It seems that Orville initiated these early enterprises. He collected bones for making fertilizer, manufactured chewing gum from sugar and tar, and launched a school newspaper, *The Midget,* with a friend. Orville also led his brother into printing, their first joint business.

Their next passion was bicycling, which had become a national obsession by the 1890s. Soon they were building, selling, and repairing bicycles—and prospering. Their bicycle shop was located across the street from their printing business, and they often used one business to promote the other.

The Wright brothers pose at their family home in Dayton, Ohio, in 1909. Orville (with mustache and argyle socks) owned the song-sheet below, a token of their fame after their first flight.

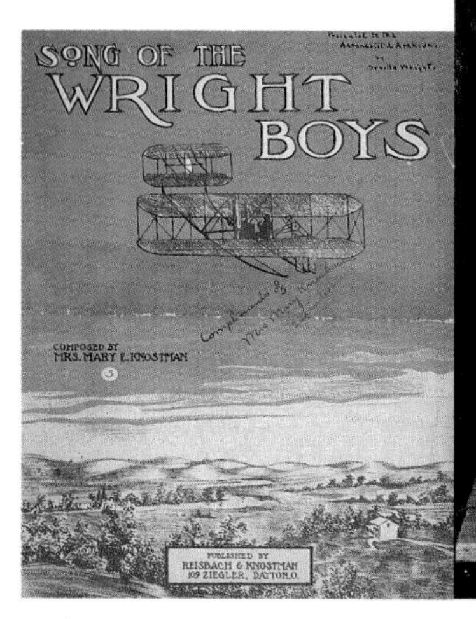

Though running two estimable enterprises, the Wrights knew they were still casting about for a sufficient challenge. As Wilbur said at the time, "I entirely agree that the boys of the Wright family are all lacking in determination and push. None of us has as yet made particular use of the talent in which he excels other men, that is why our success has been very moderate."

The Wrights' Kites and Gliders Circumstance changed this pattern. In 1896, news reached America of the death of Otto Lilienthal. The Wrights had been entranced by his glider experiments, and his death rekindled their interest in flight.

Intuitive, brilliant researchers, Wilbur and Orville Wright immediately began studying the previous four decades of experiments in flight. In addition to reading, the brothers corresponded with leading figures in aeronautics, such as Octave Chanute, and Samuel P. Langley at the Smithsonian. To ensure that he was taken seriously by Langley, Wilbur included this disclaimer in his letter: "I am an enthusiast, but not a crank . . . I wish to avail myself of all that is already known and then if possible add my mite to help on the future worker who will attain final success."

The Wright brothers test the 1900 glider as a tethered kite.

The Wrights then began applying their knowledge to the construction of a series of gliders. The first major hurdle they had to clear was creating a system for controlling and stabilizing flight. Weight shifting, the system used by Lilienthal, seemed inadequate. Another means of control had to be found.

In 1909 the Wrights' company was the largest airplane manufacturer in the world, turning out four planes a month.

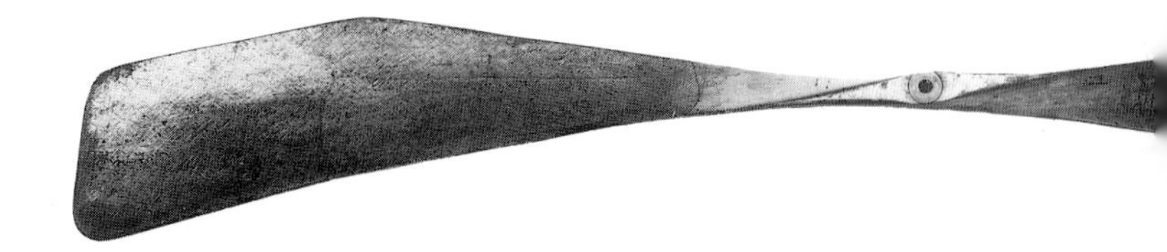

The answer they developed is called *wing warping*, a mechanism for twisting the tips of an aircraft's wings. In 1900 the Wrights tested this idea in the sand dunes of Kill Devil Hills, near Kitty Hawk, North Carolina. Their first glider, which they usually flew as an unpiloted tethered kite, cost $15 in materials. The human cost of persevering, however, was higher. The Wrights' shack in the dunes was cold, sand blew in between the boards, and mosquitoes were a constant irritation.

Initial tests proved disappointing. The glider provided less lift and more drag than they had expected on the basis of Lilienthal's data. A larger, improved glider the brothers flew in 1901 covered

The Wright brothers practiced flying gliders before attempting powered flight. This experience gave them insight into the problems of flight control. Here, Wilbur prepares to fly the 1901 glider.

The Wrights' propeller (below) represented a break with accepted ideas about propulsion—it was shaped like a twisted wing, unlike the screw-shaped device used by most inventors.

The Wrights built this wind tunnel in 1901. The fan forced a stream of air through the 6-foot tunnel at a speed of up to 27 miles per hour. The performance of the small wing shapes inside was studied through the glass panel on top of the tunnel.

315 feet in 19 seconds, at 13 miles per hour, but it still did not meet their expectations.

By the end of the 1901 test season, the Wrights were on the verge of giving up. Chanute urged them to persevere. With his encouragement, they returned to their experiments. They discovered that the data from earlier inventors could not always be trusted. Using hand-built wind tunnels, the first of which was made from an old starch box, the Wrights conducted tests to determine the best possible shape for a wing.

Using their new data, they built a biplane glider with a 32-foot wingspan in 1902. By September, they had made 50 glides and had had their first accident. Orville's crash from an altitude of about 30 feet inspired them to link the glider's wing-warping controls to a movable rudder. They also installed forward elevators to control the glider's ascent and descent.

The Wrights' Engine and Propeller The next leap forward was finding a solution to the problem that had engrossed Langley: propulsion for a piloted craft. The Wrights' team solved this problem by building their own 12-horsepower engine. Just what the engine would propel was another question.

After finding that existing propellers, including steamship propellers, were not adaptable for aircraft use, the Wrights' team invented the first operational aircraft propeller. They deduced that it was nothing more than a rotating, twisted wing, and designed a propeller shaped like a rotary wing rather than a ship's screw. Thrust was generated as the propellers spun, powered by the engine.

By 1903 the Wrights had created a craft ready for piloted, powered flight. On December 14, 1903, they flipped a coin for the first chance to fly it. Wilbur won, but he stalled and hit the ground, damaging the rudder. Three days later, Orville climbed aboard.

Kitty Hawk
December 17, 1903

Harsh winds, whipping sand, and frozen puddles ushered in the dawn of December 17, 1903. Orville and Wilbur Wright warmed themselves by a small stove in their drafty cabin, then got dressed in the starched shirts, ties, and dark suits that they wore every day. Then they waited for the weather to calm.

By 10:00 A.M., the wind showed no sign of letting up. The Wrights decided to test the Flyer anyway. They hoisted a signal banner to alert the men at Kill Devil Hills Life Saving Station. As they installed the 60- stretch of sand 100 feet north of camp, they frequently ran back inside to warm up.

By 10:30 A.M., several lifesavers had arrived, and the Wright Flyer was facing into the wind atop the launch rail. The brothers poured gasoline into the four-cylinder engine, placed the battery on the wing, and hooked its wires to the engine.

Orville ran the motor to warm it up; then he and Wilbur stepped away briefly to talk. One lifesaver remembered, "We couldn't help notice how they held on to each other's hand . . . like two folks parting

other again." Orville later recalled how risky it had been to try out a new flying machine in such blustery weather.

Then Orville positioned himself prone on the lower wing, next to the engine. He checked his controls, moving the elevator lever up and down and shifting his weight to operate the wing-warping mechanism and the rudder.

After stationing one of the lifesavers with a camera and tripod to capture the moment, Wilbur urged the spectators "not to look too sad . . . laugh and holler and clap . . . try to cheer Orville up."

At 10:35, Orville released the wires holding the Flyer to the rail. The craft moved slowly down the track as Wilbur ran alongside, balancing the wing. Near the end of the track Wilbur let go, and the plane lifted into the air. For twelve seconds the Wright Flyer flew erratically, climbing and descending, then suddenly skidded onto the sand 120 feet from where it had taken off.

An hour later, Wilbur took a turn and flew 195 feet. Then Orville flew again, this time covering 200 feet. Later, Wilbur's second try accomplished a flight of 852 feet in 59 seconds.

Though the control mechanisms were not perfect, it was clear the Wright Flyer had succeeded. A young boy who had witnessed the event darted down the beach toward Kitty Hawk, shouting, "They done it! They done it! Damned if they ain't flew!"

While the Wrights were discussing plans to repair some damage to the aircraft caused on Wilbur's second flight, the wind overtook the Flyer and tossed it end over end. The wing structures were broken, the engine damaged, and the chain guides bent. There would be no more flights that year, and in fact the Wright 1903 Flyer never flew again.

Orville and Wilbur telegraphed the following message to their father, Bishop Milton Wright:

THE WESTERN UNION TELEGRAPH COMPANY.
INCORPORATED
23,000 OFFICES IN AMERICA. CABLE SERVICE TO ALL THE WORLD.
ROBERT C. CLOWRY, President and General Manager.

RECEIVED at

176 C KA CS 33 Paid. Via Norfolk Va

Kitty Hawk N C Dec 17

Bishop M Wright

7 Hawthorne St

Success four flights thursday morning all against twenty one mile wind started from Level with engine power alone average speed through air thirty one miles longest 57 seconds inform Press home Christmas . Orevelle Wright 525P

Wright 1903 Flyer

The problems that had bedeviled other experimenters in flight—lift, flight control, and adequate propulsion—were at last resolved by the Wrights. Each mechanism of the 1903 Flyer shows their ability to ignore common assumptions about flight and to develop and perfect a workable flying machine.

The Flyer's wingspan measures 40 feet, 4 inches. The craft weighs 605 pounds.

Muslin-covered wings are curved to provide greater lift. The trailing edges of wings can be warped to bank the airplane. Twisting the left wing down and the right wing up produces a roll toward the right. Banking stabilizes the plane during turns.

Elevators—panels fixed to the front of the Flyer—are moved up and down to raise and lower the aircraft's nose.

The Flyer is powered by a four-cylinder, water-cooled engine. Eight-foot propellers, attached to the engine with bicycle chains, are mounted behind the Flyer's wings to push it forward. The propellers are shaped like twisted wings.
The rudder assists in steering the Flyer.
The elevator lever is operated by the pilot's left hand.
Flexible cables link rudder to wing-warping controls attached to the pilot's hip saddle.
Spruce struts support the wings.

The Wrights collaborated with a local machinist, Charles Taylor, to build an engine for the Flyer. The water-cooled engine weighed 200 pounds and was capable of producing about 12 horsepower, spinning the propellers up to 350 revolutions per minute.

Publicity—and Patent Conflicts The press greeted the Wrights' historic news with disbelieving silence or small, inaccurate stories on the back page. Langley's much-publicized failure had made skeptics of the public and the press. As the local representative of the Associated Press said, "Fifty-seven seconds, hey? If it had been fifty-seven minutes then it might have been a news item."

Undaunted, the Wrights moved their experiments closer to home, to Huffman Prairie, a field near Dayton. Here they developed the Wright Flyer II, which flew the first circle in the air.

As they continued to improve the Flyer, the brothers became preoccupied with protecting their patents and selling their invention to the army. They were convinced they had an unbeatable lead in producing functional aircraft.

The flight of rival inventor Glenn Curtiss in 1908 was shocking proof that the Wrights were wrong. They quickly signed a contract with the army for far less than their asking price. From that time forward, the Wrights engaged in exhausting patent wars with Curtiss and other American inventors. The conflict was finally stilled by World War I, when patriotic duty forced a cross-licensing agreement between the competing companies.

Belated acclaim was finally bestowed on the Wrights in 1908 when Wilbur traveled to Europe and amazed aviators, inventors, and the public with his flying skill and the maneuverability of the Wright A biplane. The Wright brothers and their sister Katharine were feted as the first great celebrities of the new century. Their fame followed them home, where President William Howard Taft presented them with gold medals from the U.S. Congress, the Smithsonian Institution, and the Aero Club of America. Wilbur died in 1912, but Orville lived a long life—by the time of his death in 1948, he had seen many more astounding developments in aviation.

Wilbur Wright flies in France in 1908. Fame did not change his reserved and unassuming manner. Declining to make a speech at a banquet in France, he quipped, "I know of only one bird, the parrot, that talks, and it doesn't fly very high."

The distance flown by the Wright brothers in their first flight was shorter than the wingspan of a Boeing 747.

Glenn Curtiss

Like the Wrights, Glenn Curtiss began his career by building bicycles, but by 1901 he was building engines and racing motorcycles. After the Wrights rejected his offer to build custom engines for their planes, Curtiss joined forces with Alexander Graham Bell and his Aerial Experiment Association (AEA).

Called "Bell's boys," the members of AEA launched their first plane on Keuka Lake in New York, in March 1908. The piloted biplane "sped over the ice like a scared rabbit for two or three hundred feet, and then, much to our joy, it jumped into the air." The AEA's first flight covered 319 feet.

By June of the same year, though, the AEA was ready for a bigger challenge: *Scientific American*'s prize for the first American to make a public flight over a measured course. The publisher had set up the contest as an easy win for the Wrights, but their refusal to compete left the field wide open.

Flying in the *June Bug*, Curtiss covered a distance of 5,090 feet in less than two minutes and became an instant hero. Less than a year later, he left the AEA to form an aircraft manufacturing company with his colleague Augustus Herring.

In 1909, Curtiss took one of his biplanes, the *Rheims Racer,* to the air competition in Rheims, France. With a single airplane and an untried engine, Curtiss faced French aviators with spare airplanes, huge crews, and tons of equipment.

Against these odds, Curtiss entered only a single race: the Gordon Bennett Trophy for speed. In a practice run, he was battered by so much turbulence that he swore he would not go back up. His tune changed when he heard that his time was the fastest of the meet. In his bid for the cup, Curtiss flew into the turbulence with his throttle wide open. His *Rheims Racer* won by six seconds.

Glenn Curtiss set speed records for airplanes and motorcycles, directed the AEA's experimental work, and developed aircraft, motors, and speedboats. Involved in bitter patent battles with the Wrights, he tried to discredit their claims by "proving" that Langley's Aerodrome was airworthy—a scheme calling for major modifications to the Aerodrome.

FEDERATION AERONAUTIQUE INTERNATIONALE

AERO CLUB OF AMERICA

No. 1

The above-named Club, recognized by the Federation Aeronautique Internationale, as the governing authority for the United States of America, certifies that

Glenn H. Curtiss

born 21st day of May 1878 having fulfilled all the conditions required by the Federation Aeronautique Internationale, is hereby licensed as Aviator.

Dated June 8th, 1911.

Allan A. Ryan, President.

G. F. Campbell Wood, Secretary.

[SEAL]

Signature of Licensee: Glenn H. Curtiss

Pioneer aviator Glenn Martin founded one of the first airplane factories in the United States, in 1909. During the two world wars his company specialized in producing bombers.

Glenn Martin

The first airplane Glenn Martin caught a glimpse of was a Curtiss biplane forced down in Santa Ana, California, with engine failure. Long interested in flight, he decided at that moment that pursuing flight would be his future. Using his expertise with automobiles and drawing on his memory of the downed craft, Martin built his first plane in 1907.

Fashioned from wood, wire, steel tubing, and cloth, Martin's first airplane never flew. Undismayed, he developed a new design, incorporating the best features from planes built by both Curtiss and the Wrights and adding his own innovations. He built his second airplane in a church and then tore the doors down to get it out. Martin's first flight occurred in August 1909 in a 160-acre lima bean patch. He covered 100 feet and never rose more than 8 feet off the ground.

From that point, Martin steadily improved the performance of his aircraft and his ability as a pilot. As a well-paid stunt flier, Martin proved so entertaining that the Aero Club of America awarded him an Expert Aviator certificate (Curtiss had received the first one awarded.)

Martin saved his earnings as a stunt flier to fund his own company. One of his dreams was to tap the military market. In a display designed to attract this business, he bombed a mock fort from the air with bags of flour while an assistant set off black powder explosions on the ground.

Martin's first military contract, though, did not result from such grandstanding. In 1913, the army's San Diego pilot training center was using outdated planes, in which many students were crashing. Martin quickly created a trainer out of a biplane he had been designing for parachute drops. It was driven by a 90-horsepower Curtiss engine and featured an enclosed fuselage with two cockpits. With the new airplane, the army's casualty rate dropped dramatically, and the Glenn Martin Company had built the world's first successful flight trainer.

Flight Control

Unlike automobiles and ships, airplanes do not travel on a flat surface. To keep an airplane stable in the air, the pilot must be able to control movement on three axes: pitch (pointing the nose up or down), roll (the banking of the wings), and yaw (horizontal movement to the left or right).

Pitch
Pitch is controlled by an elevator, which points the nose of the plane up or down. Elevators are usually part of the tail assembly.

International Competitors

Six years after Kitty Hawk, the United States no longer dominated the field of aviation. European designers and pilots completely outclassed the Americans at the first international air meet, held in Rheims, France, in August 1909. Even though Glenn Curtiss won the prize for speed, all other prizes went to the Europeans. Of the 22 participating aviators, most were French.

Though the French may have missed the distinction of achieving the first powered, piloted flight, they embraced the fledgling field of aviation with unmatched zeal. One month after Rheims, an international exhibition opened in Paris. Of the 333 aircraft exhibitors in 10 categories, 318 were French. And by 1911 the world's speed, endurance, and altitude records were held by French pilots flying French planes.

Americans lost the lead as the Wrights devoted themselves singlemindedly to securing patents, which halted their own research and limited the work of other inventors. The Europeans, in contrast, blatantly borrowed ideas from each other and from U.S. inventors.

Social structure also played a role. The aristocrats of Europe became champions of aviation, and they had access to officials holding government purse strings. American aviators lacked such

Crashes, like this one at the Rheims air meet, were a constant risk in the fragile airplanes of the time.

Roll
Roll is controlled by ailerons—flaps on the trailing edge of a wing that cause one wing to rise while the other tilts downward.

Yaw
Yaw is controlled by a rudder on the tail of the plane. The rudder steers the plane to the left or right. The ailerons and the rudder work together to turn and bank the plane.

connections, and money to support innovation was harder to find. In the years leading up to World War I, the French alone spent more than $35 million on flying machines, whereas the U.S. government appropriated a mere $685,000.

Alberto Santos-Dumont

The Europeans' steady progress was punctuated by three notable events. The first occurred on November 12, 1906, when Alberto Santos-Dumont became the first European to achieve sustained flight in a powered airplane. A Brazilian living in Paris, Santos-Dumont was known for landing balloons on the sidewalk of the Champs-Elysées and going upstairs to his apartment for coffee.

Like many self-taught early aviators, Santos-Dumont learned quickly. In September 1906 a plane of his own design hopped less than 50 feet and crash-landed on its tail. One short month later, he remained aloft for seven seconds and traveled 197 feet. And in November, with other aviators eagerly watching, Santos-Dumont flew 722 feet in 21 seconds. This flight gained him the French Aero Club's prize for the first flight covering more than 100 meters.

Alberto Santos-Dumont was an avid experimenter and self-promoter. His balloons, gliders, airships, and airplanes delighted Parisians and greatly increased public interest in aviation.

Farman biplanes set several height and distance records between 1909 and 1911. Brothers Henri and Maurice Farman built some of the first airplanes to use ailerons instead of wing warping.

Henri Farman

Then, on January 13, 1908, the Archdeacon-Deutsch prize, established in 1904 to recognize the first recorded powered flight of a 1-kilometer circle, was finally claimed. Henri Farman, son of an Englishman living in Paris, made the attempt in a modified biplane he had recently purchased from Gabriel Voisin, an up-and-coming aircraft designer. He took off, flew around a pylon 500 meters away, and returned to the point of his departure. Never rising above 40 feet, Farman crossed the finish line 12 feet off the ground.

Hailing Farman as a hero, the London *Times* falsely claimed that "nothing of the kind has ever been accomplished before." In fact, the Wrights had done far better than either Santos-Dumont or Farman years earlier.

Other Designers

Three inventors who put planes in the air between 1909 and 1914 had a particular influence on aircraft used later during World War I. One was Geoffrey de Havilland, an English designer whose craft were purchased by the British government. By 1911, he built the B. E. I biplane, the forerunner of the classic single-seated scout and fighter planes soon to be developed.

Anthony Fokker, the son of a Dutch coffee planter in Java, also became closely associated with the development of fighter planes. An engineer and demonstration pilot with a performer's panache, Fokker did not hesitate to borrow ideas, even

The Blériot XI was produced in France for both the civilian and the military markets. The first plane to fly across the English Channel, in 1909, the monoplane's fabric-covered wings spanned a mere 25 feet.

buying junked planes to check them over. By 1913 he had won key contracts from the German army.

Finally, Russia's Igor Sikorsky was inspired by the Wright brothers to build and fly his own designs. After building several successful single-engine aircraft, he concentrated on large, multi-engine planes. His four-engine biplanes set many records and were used effectively by the Russian army in World War I.

Louis Blériot flew the English Channel without a compass or any other navigational instrument. France, Blériot's home base, was a leader in the fledgling field of aviation, and many British enthusiasts went there to train as pilots.

Blériot's Channel Crossing

The next prize to be taken was truly a world first. The London *Daily Mail* offered a prize of £1,000 for the first flight over the English Channel made between sunrise and sunset. One of the leading contenders for the prize was Louis Blériot. He took off in his Blériot XI monoplane at 4:40 A.M. on July 25, 1909, from the coast of France.

Caught in mist as he approached the English coastline, Blériot spotted several ships and followed them to the Dover cliffs as the sailors cheered. Fighting the wind, Blériot cut his ignition 60 feet above an empty field and landed at 5:20 A.M. "I didn't point myself," he said afterward, "rather I flung myself toward the ground."

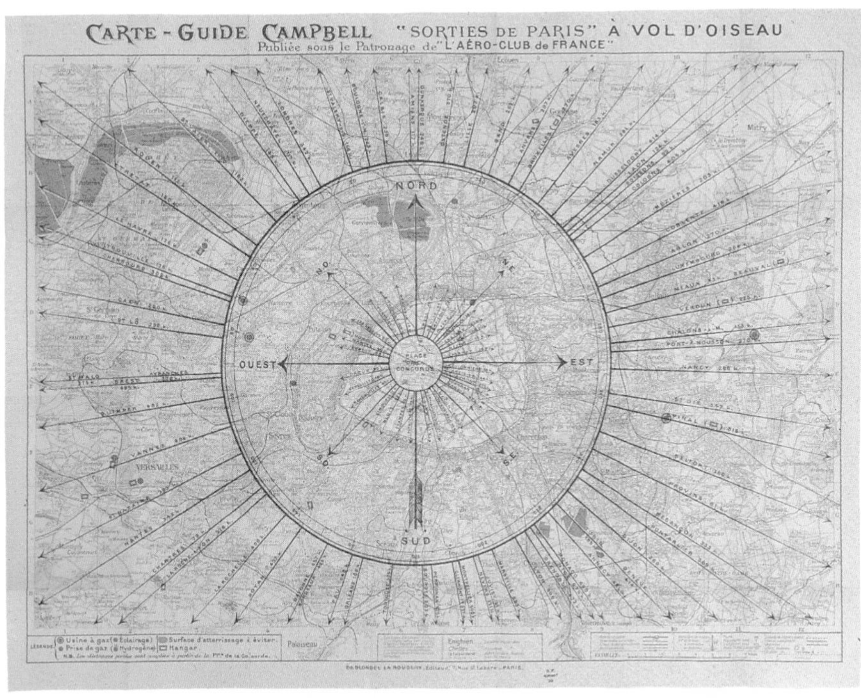

This map of 1904, detailing proposed aviation routes around Paris, reflects France's enthusiasm for flight.

World War I: The Airplane and Modern Warfare

Aviation in World War I began with friendly greetings between reconnaissance pilots of opposing sides but soon led to deadly combat and the development of weapons for aerial warfare.

The mobility of fliers in World War I seemed a bitter contrast to the static, demoralizing war in the trenches. But air power created a new dimension of destruction, in which civilians and soldiers alike faced strafing and bombing attacks, and the defensive boundaries of oceans and mountains easily gave way to an enemy's fighters and bombers.

Introduced in 1917, the Fokker Dr. 1 Triplane was flown by Germany's leading ace, Manfred von Richthofen, in the last months of his career. He scored 19 of his 80 victories flying this model.

Often referred to as "chicken coops," France's Voisin airplanes filled many roles in World War I. The Voisin VIII (above) was used to bomb German towns from 1915 to 1917.

Air Power at the Beginning of World War I

In August 1914 the world hardly seemed ready to deploy aircraft in battle. Despite Europe's enthusiasm for flying, little had been done to train aviators or to build aircraft suited to combat. Armies on both sides of the conflict fumbled with the process of organizing aviation branches, and there were not enough planes and pilots in all of Europe combined to deserve the term "air force."

In 1914 France and Germany could mobilize only a few hundred planes each, about half of them ready for combat. Germany's air force of approximately 250 planes was dominated by small, single-wing Taubes and Albatros biplanes. France had the advantage in number of pilots, due to the aviation craze that had swept the country, but most of them flew obsolete Farmans and Blériots. The British could muster only about 180 aircraft—an assortment of B.E.2s, B.E.8s, Avros, and others. Less than half of them were ready for active duty.

The mobilized planes were flimsy machines whose engines had a disconcerting tendency to stop during flight—if they started at all. Many of the pilots who flew the British planes to France wore inner tubes around their waists in case they had to ditch in the English Channel. The best planes of the time could manage 6,500 to 9,000 feet in altitude, 60 to 70 miles per hour, and 200 miles in range. On the ground, the planes required constant attention and overhaul—even when new. Pilots complained of parts that were missing or even installed upside down.

Reconnaissance at the Marne, 1914

On August 22, 1914, General John French of the British Expeditionary Force (BEF) in France got a surprise: reconnaissance pilots reported that the German army, which had begun a wide sweep across Belgium to reach Paris from the west, had changed direction. It was taking a shorter southern route toward Paris—right toward the BEF.

If the Germans had surprised the BEF, they could have made a direct push toward Paris. Instead, the Allies were able to solidify their position at the Marne River and, in September, hold off the German advance. General French praised the Royal Flying Corps pilots for their timely information, which helped prevent a disaster for the Allies.

Aerial photography was a revolutionary method of gathering information during wartime. Photographs such as the one above were used to make accurate maps for planning military strategy. Troop movements and the stockpiling of arms and supplies could also be documented.

To take an aerial photograph in World War I, an airman had to lean over the edge of the plane (without a parachute), snap the photograph, and change the plate before taking another picture. The camera was attached to the side of the plane or, early in the war, was strapped to the aviator's chest.

Reconnaissance Flights

Because the airplanes were unarmed, they could not be used for offensive purposes. Their first mission was reconnaissance. Pilots reported on enemy positions, the movement of reinforcements and supplies, and the location of munitions dumps. They spotted for field guns and helped soldiers in the trenches direct their fire accurately by sending messages from the planes by signal lamp. In spite of the ragtag nature of their equipment, the fliers performed their jobs well.

Most of the time, British and French leaders believed and acted upon information from the air services; Allied armies changed position in response to reports of German troop movements. By some accounts, had it not been for information gathered by aircraft, the Germans could have taken Paris in the first few weeks of the conflict.

"Winning your wings" is the major goal for pilots in training, whether for airplanes, airships, or even balloons.

By 1915 aerial reconnaissance had become more sophisticated. Telegraphy had replaced the dropping of messages and hand signaling. Cameras captured explicit detail from the air. By the time of the Battle of Neuve-Chapelle in March 1915, all maps for the engagement were based on aerial photos.

At first, reconnaissance pilots from opposing armies waved as they flew past each other, but inevitably they began trying to keep each other from gathering information. Pilots took pot shots at each other with guns they carried under their heavy flight suits. But these random methods proved ineffectual—the massive firepower produced by machine guns was needed. Conflict in the air escalated as airplane designers transformed airplanes into weapons of war.

Balloons and Reconnaissance

Tethered, hydrogen-filled balloons were used during World War I by both German and Allied troops to observe enemy forces. Observers in wicker baskets were suspended beneath the balloon, where they directed artillery fire by telephone. Attached to the ground by sturdy cables, balloons were quickly lowered when enemy pilots tried to shoot them down. Well defended by ground guns, they were dangerous and difficult targets.

Cigar-shaped airships called dirigibles or Zeppelins also were used for long-range reconnaissance and bombing. They required the assistance of large ground crews in launching and landing.

A German reconnaissance balloon is slung with a gondola for the observers, who bailed out swiftly if the balloon came under enemy attack.

Dirigibles and World War I

Because they flew at high altitudes, temperatures within rigid airships were rarely above freezing. Crews wore fur-lined coats and thick mittens and received a ration of brandy to stave off the cold.

More than 640 feet long, the German Zeppelins were a frightening sight, looming in broad daylight over the English countryside. Though the physical damage caused by the bombs they dropped was often slight, the psychological effect on the British was profound.

By 1914, the British had only one single dirigible, which broke in two before its first ascent. In contrast, as early as 1910 German advisers were working with Count Ferdinand von Zeppelin on a special dirigible capable of carrying a load of bombs across the channel to England. By 1915 the Germans were regularly sending Zeppelins on bombing missions over England.

Germany's dirigible attacks peaked in 1916. By then, pilots of British fighter aircraft had become skilled in flying above the hydrogen-filled airships and shooting them down with incendiary bullets. The cost in men and materials was simply too high.

The heaviest attack occurred on September 2, 1916, and involved 16 airships carrying 32 tons of bombs. Four British citizens lost their lives; in contrast, the Germans lost 16 crew members and a brand-new airship.

By the end of the war, rigid airships were no longer used in long-range bombing attacks; both the Germans and the Allies used them mainly for reconnaissance.

The Flying Aces

The ace system was started in France to honor pilots who had scored five confirmed aerial victories. The Germans developed a similar concept requiring ten victories; they called an ace an *Oberkanone,* a "top gun."

Honor was heaped on these aerial warriors. The air war was more easily romanticized than the mud of the trenches. Still, fighting in the air was always about death. The pilots could see the faces of the men they shot down, and they knew they risked the same fate.

Max Immelmann (right), Germany's "Eagle of Lille," was famous for his deft dogfighting maneuvers. The whole country mourned when he was shot down in 1916.

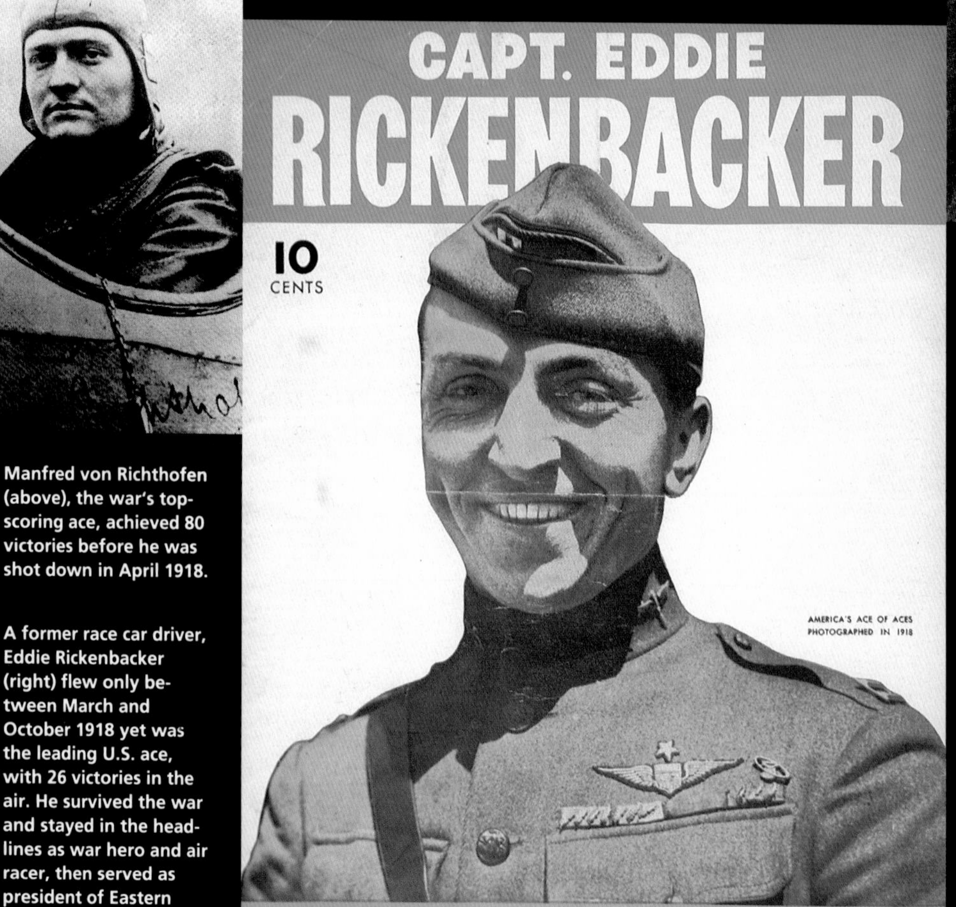

Manfred von Richthofen (above), the war's top-scoring ace, achieved 80 victories before he was shot down in April 1918.

A former race car driver, Eddie Rickenbacker (right) flew only between March and October 1918 yet was the leading U.S. ace, with 26 victories in the air. He survived the war and stayed in the headlines as war hero and air racer, then served as president of Eastern Airlines from 1938 to 1959.

Canada's top ace, Billy Bishop, scored 72 victories and survived the war.

The best-known German ace was Baron Manfred von Richthofen. Known as a cold and calculating fighter, he brought the attitudes of an aristocratic huntsman to dogfights. He maintained a collection of silver cups, each engraved with a description of one of his victims. His silversmith worked hard during Bloody April in 1917, when the Baron shot down 21 aircraft. Dubbed the "Red Baron," Richthofen usually flew a bright red Albatros or a Fokker Dr.1.

France's greatest ace was René Fonck, a master at conserving ammunition. He was frequently able to down an aircraft with only five or six rounds, placed, in his words, *comme avec la main* ("as if by hand"). On one day in 1918 he downed six planes, including three two-seaters that crashed within 400 yards of one another. French ace Charles Nungesser was wounded several times and joked that there was not a bone in his body without an aluminum clip. He managed many of his strikes when he couldn't even walk—he had to be carried to and from his airplane. Like many aces, he carried a distinctive insignia on his plane: a black heart set with a skull and crossbones, two candles, and a coffin.

Britain's most successful ace was Major Edward Mannock. Suffering from astigmatism in the left eye, he bluffed his way through his military medical exam. Mannock was a consummate flight leader: patrols under his leadership were never attacked by surprise.

The best-known American ace was Captain Edward Vernon Rickenbacker, who ignored his mother's admonitions to "fly slow and stay close to the ground." Rickenbacker first went to France as General John Pershing's chauffeur and completed advanced flying and gunnery courses on his own time. In March 1918 he flew with the first American patrol to cross over enemy lines; he shot down 26 enemy planes during the war.

Early Fighter Planes

To fight effectively, planes required machine guns. A single-seater's gun had to be mounted within reach of the pilot and easy to aim. The ideal position was on the upper fuselage, directly in front of the windscreen, but this created a problem: the stream of bullets would pass directly through the rotating propeller, splintering the wooden blades.

The first solution to this engineering challenge came from the French pilot Roland Garros, with the help of the French aircraft designer Raymond Saulnier. They placed steel plates on the backs of the propeller blades to deflect bullets that otherwise would have broken the blades. With this arrangement, a pilot simply aimed the plane at the target and fired. Garros downed three enemy planes in less than a month and mystified the Germans with his ability to shoot through the propeller. Then he was captured by the Germans, along with his plane.

The Germans copied the propeller's deflector plates, but their prototype did not work. The force of their machine gun shot the blades off the propeller. The German high command then called on 25-year-old aircraft designer Anthony Fokker. He scrapped the French system and built a device that synchronized the propeller and the gun. Within 48 hours he had a successful prototype: it briefly interrupted the fire of the machine gun whenever the propeller blades passed in front of the gun.

With this mechanism mounted on the Fokker Eindecker E.III monoplane, the air war became serious indeed.

The Fokker Eindecker was introduced on the western front in the autumn of 1915 and dominated the skies until the summer of 1916.

Members of the Lafayette Escadrille pose with their mascot lion cubs.

The Lafayette Escadrille

Restless with U.S. neutrality, some American pilots served as volunteers with British and French forces. The Lafayette Escadrille, the much romanticized, elite fighter squadron, was composed entirely of U.S. volunteers. A month after its organization in April 1916, the Lafayette Escadrille entered combat in the Battle of Verdun as squadron N-124.

W. K. Vanderbilt headed the U.S. committee that financially backed the Lafayette Escadrille and provided them with mascots—lion cubs named "Whiskey" and "Soda."

The squadron (with aces like Raoul Lufbery, William Thaw, Norman Prince, and James N. Hall) scored 39 victories against the Germans—but nearly a third of the unit died in the effort.

War in the Air

Germany Dominates

Though its scramble into the air had begun somewhat awkwardly, Germany gained dominance in the skies at the outset of the war. Germans flying German planes already held both the world endurance record (24 hours, 12 minutes) and distance record (1,178 miles). Now the German high command took the aerial offensive, dispatching Taubes over Paris and dirigibles over London on bombing missions.

The Eindecker Germany truly took control of the skies with the feared Eindecker E.III monoplane. This aircraft's design was derived from a French plane, one of the Morane-Saulniers. It was a tiny plane, with a wingspan of only 28 feet. Its 80-horsepower engine, also a copy of a French design, gave the plane a top speed of 82 miles per hour. Equipped with a synchronized machine gun in the spring of 1915, it was called the Eindecker ("single wing")—a simply unbeatable fighter. For the next year, Allied pilots ruefully described their own planes as "Fokker fodder." Germany's invincibility in the air became known as the Fokker Scourge.

Early in the war, armies dispatched cavalry units to observe enemy movements. By 1915, the airplane had virtually replaced the horse-mounted observers.

Germany's ace pilots, such as Max Immelmann and Oswald Boelcke, personified Germany's conquest of the air. Feted as heroes, they trained new pilots and created and perfected the maneuvers and strategies of successful aerial fighting. Together with the Eindeckers, they ruled the skies from October 1915 until the middle of 1916.

The Allies Respond

The D.H.2 The introduction of new Allied designs, however, challenged Germany's control of the air. British designer Geoffrey de Havilland contributed the F.E.2b and the D.H.2. The D.H.2 was a particularly impressive machine. With a speed of 90 miles per hour and a Lewis machine gun mounted in front of the pilot, it could match the Fokker Eindecker.

Another British craft, the Vickers FB.5 Gun Bus, was the first plane to be grouped into a homogeneous unit, the No. 11 Squadron of the Royal Flying Corps (RFC). The squadron left for France on July 25, 1915. Such units soon became common on both sides. By the end of the year, the Germans had deployed more than 80 individual sections of aircraft, including special bombing squadrons. These sections later evolved into the *Jagdstaffeln* (fighter groups).

The Nieuport 11 France's answer to the Eindecker was the highly maneuverable Nieuport 11, called the *Bébé*. Like the D.H.2, the Nieuport 11 was a fast plane with a Lewis gun mounted above the wing. Together with the British planes and the other Nieuport models, the *Bébé* bolstered the Allies' strength in the air. Germany was further set back on June 18, 1916, when German ace Max Immelmann, the "Eagle of Lille," was shot down in his Fokker Eindecker by an F.E.2b. The SPAD VII, which entered combat in the fall of 1916, also challenged the Germans.

The fast and maneuverable Nieuport 11 successfully challenged the Fokker Eindecker. A sesquiplane (with a full upper wing and shorter lower wing), it was armed with a Lewis gun.

Fighter Planes Later in the War

Throughout the last two years of the war, the Allies and the Central Powers vied for advantage in the air. In 1917, the revolution in Russia ended the war on the eastern front and brought additional German troops to the western front. At the same time, the arrival of U.S. forces brought hope to the Allies. Still, the trench lines of the western front barely moved. Aircraft designers strove to meet the challenges of each new airplane developed by the other side, hoping to end the stalemate on the ground by gaining superior power in the air.

A reliable single-seat fighter introduced in 1917, the Sopwith Camel, was a bit difficult to handle but a real workhorse. Armed with two forward-firing guns, the Camel could also be equipped with racks to carry a load of four small bombs. The Camel scored more victories against German airplanes than any other Allied plane in World War I.

SPAD XIII and the Albatros D.Va

The SPAD XIII, flown during the last year of World War I, proved itself a sturdy biplane fighter. Built by the French manufacturer SPAD (which stands for *Societé Pour l'Aviation et ses Derivés*), it succeeded the SPAD VII, which was less maneuverable and carried only one machine gun. Many U.S. pilots flew SPAD XIIIs.

The SPAD XIII's 220-horsepower Hispano-Suiza ("Hisso") engine provided more power than the engines of Germany's Albatros fighters, but it was not always reliable—the only real weakness of this famous fighter.

Two synchronized Spandau machine guns fire through the propellers.

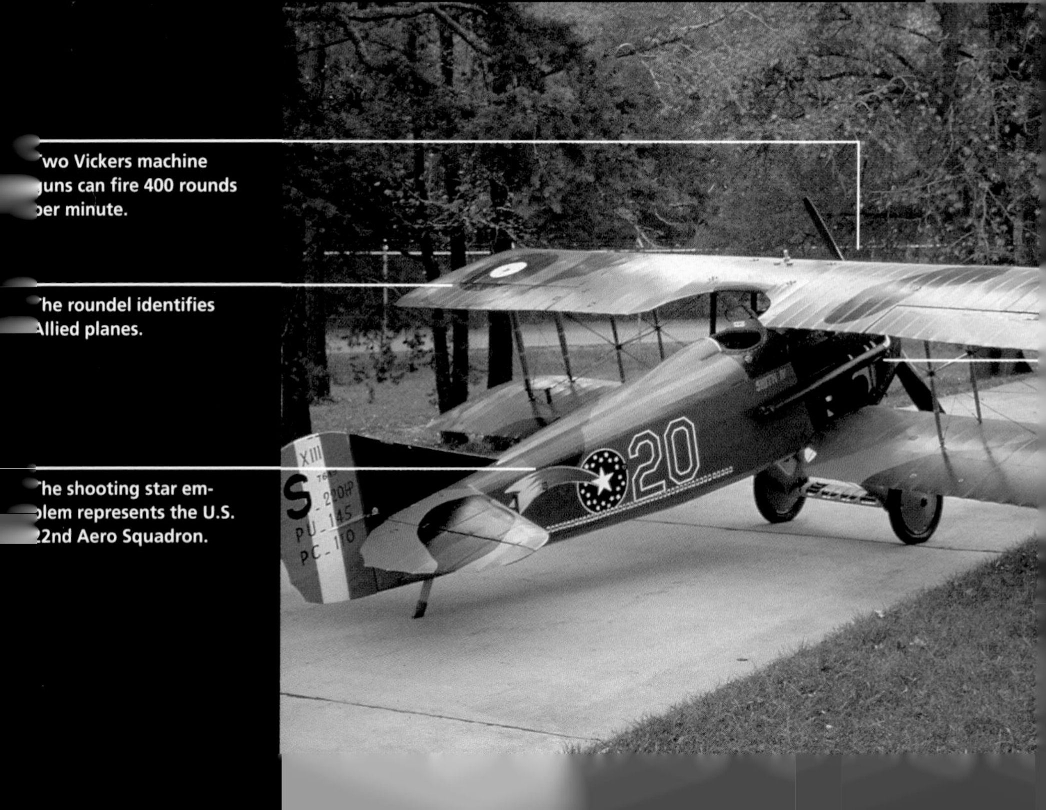

Two Vickers machine guns can fire 400 rounds per minute.

The roundel identifies Allied planes.

The shooting star emblem represents the U.S. 22nd Aero Squadron.

The Mercedes engine delivers 180 horse-power, giving it a maximum speed of over 115 miles per hour and an altitude of over 20,000 feet.

Ailerons bank the plane for making turns.

Yellow and green tail stripes represent Germany's Jasta 46 fighter squadron.

Fabric-covered wings are painted with Germany's cross pattée.

Manufactured between 1916 and 1918, Germany's Albatros biplane fighters were dependable and easy to fly, though generally outperformed by the Allies' Sopwiths and SPADs.

The D.Va, the final model in the line, was introduced on the front lines in mid-1917. Along with the Albatros D.V, it was flown by many aces, including Hermann Göring, Ernst Udet, and Manfred von Richthofen (who won most of his victories in an Albatros).

A 200-horsepower Hispano-Suiza engine gives a maximum speed of 130 miles per hour and an altitude of over 22,000 feet.

The Nieuport 28 entered service in 1917. It had a maximum speed of 120 miles per hour and was armed with two Vickers guns. Many U.S. pilots flew the Nieuport 28—the hat-in-the-ring emblem on the fuselage signifies the U.S. 94th Pursuit Squadron.

The Fokker D VII The superior design of the Fokker D VII set it apart from other planes over the battlefields of World War I. This biplane had thick, almost cantilever wings of wood and fabric with a fuselage constructed from welded steel tubes. Its 185-horsepower BMW engine delivered a top speed of 120 miles per hour and could power the plane to a height of 23,000 feet. The Fokker D VII, as the standard fighter of German squadrons at the end of the war, was the only aircraft the Armistice agreement specifically demanded be turned over to the Allies.

The Sopwith Snipe The Sopwith Snipe went into production only a few months before the war's end. It saw limited action on the western front, where it was flown by three squadrons and employed as a more powerful replacement for the Sopwith Camel. A pleasure to fly, it was the RAF's standard fighter until the mid 1920s.

World War I Bombers

Early Efforts

The first aerial bombs dropped during the war were not the exploding kind. Between August and October 1914, the Germans flew over Paris in Taube monoplanes, dropping weighted messages such as this: "The German army is at the gates of Paris. There is nothing for you to do but surrender."

Early in the war, fighter pilots dropped incendiaries and explosives as well as propaganda. They had no bombsights on their planes, so luck rather than skill determined the outcome. Often bombing raids were merely secondary duties for fighter pilots. On August 29, 1914, for example, a German fighter flying over an RFC base in France dropped three bombs, with no effect. On September 1, a British pilot on reconnaissance tossed a bomb at enemy cavalry standing at a crossroads. He claimed that it scattered the forces considerably.

Bombing Raids

The war's first real bombing raid occurred on October 8, 1914, when two British Sopwith Tabloids flew to Antwerp, carrying 420-pound bombs. Their orders were to attack the dirigible sheds at Cologne and Düsseldorf. One pilot, lost in heavy mist, dropped his bombs on the main railroad station but caused little damage. The second pilot found the target and managed to destroy a new Zeppelin.

By the end of 1914, German airplanes were dropping small bombs on British soil. The Germans' interest in bombers, however, was slowed by their commitment to Zeppelins. By early

The graceful Etrich Taube monoplane was Germany's most widely used aircraft at the beginning of World War I. Used chiefly for reconnaissance, it also flew on bombing missions until 1916.

Bomber Design: 1917

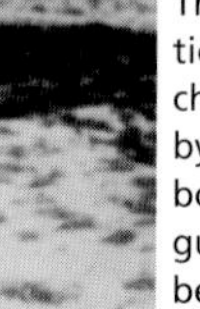

1915, German Zeppelins were conducting bombing raids on Britain, which did more psychological than physical damage. The Allies, on the other hand, never envisioned using airships in this way. As a result, they concentrated on designing true bombers.

The D.H.4 The British D.H.4 bomber, first deployed in 1917 and used as a day bomber, became the mainstay of British and American forces. D.H.4s were flown in a tight formation of six or twelve planes; their pilots covered for each other, using defensive fire when needed. Formation flying was challenging—maintaining position while carrying a heavy bomb load (two 230-pound or four 112-pound bombs, hung from the lower wing) was difficult for the less experienced pilots, who often lagged behind—easy targets for the Germans.

Other British bombers included the impressive Handley Page 0/100, used for night bombing, and the F.E.2b.

The Gothas By 1917, the Germans, discouraged by the high failure rate of the Zeppelins, were also building specialized heavy bombers. One of the best known was the Gotha G V, which launched daylight raids on London in July 1917, then later switched to night attacks. Over the course of the war, German aircraft dropped approximately 9,000 bombs on England—a total of 280 tons. Bombing raids made the term *civilian noncombatant* obsolete, and the phrase "the bomber will get through" summed up war in the air—no person, no location was safe from weapons with wings.

The Zeppelin-Staaken R VI bomber was enormous—its 138-foot wingspan was almost twice that of the Gotha bombers. Its crew included two pilots, a navigator, two mechanics, a fuel attendant, and a radio operator.

The Caproni Ca-36 bomber, the definitive model of the Caproni Ca 3 series, was manufactured in Italy toward the end of the war. One of the most effective Allied bombers, it carried a crew of four and up to 1,000 pounds of bombs.

The Curtiss "Jenny"

U.S. manufacturers turned out European-designed fighters and bombers (only the D.H.4 saw frontline service) for the Allies. They added the Liberty engine to some of them but actually designed only one World War I classic: the Curtiss JN-4D, known as the Jenny. It was used extensively—and exclusively—as a trainer. Many American pilots learned to fly the Jenny in Texas before going on to fight in Europe. After the war, the U.S. Army sold surplus Jennies at a bargain price, outfitting pilots for the barnstorming years that followed.

Goggles were essential gear for pilots flying World War I fighters. The open cockpits were constantly blasted by freezing wind, and bundling up in leather and wool was necessary even on warm days. The roar of engines and the smell of exhaust fumes also tested the pilots' endurance.

Bombers and a New Kind of War

Vivid proof that military aircraft had come into their own was the Allied attack on the German salient at Saint Mihiel in September 1918. Led by the U.S. Air Service's Colonel William "Billy" Mitchell, the attack force of 1,500 fighters and bombers was the largest armada of planes ever organized. Mitchell dispatched 500 fighters and light bombers over the front lines, strafing and bombing. Two more waves, of 500 planes each, slammed the German rear guard, destroying supplies, communications, and transportation routes.

Toward the end of the war, both Allies and Germans were bombing military targets night and day. They attacked railheads, ammunition dumps, and airfields. Both made the obligatory accusations concerning their opponent's attacks on "plainly marked hospitals."

The Allies and the Central Powers learned different lessons about bombing in World War I. The Germans became convinced that bombing raids were not worth their high cost and concentrated on other uses of their aerial forces. The British, however, were still licking their wounds after the bombing attacks on London; they vowed to build a bomber that could reach the German heartland.

Despite the horrors of the war, aviation was still a young, 15-year-old technology. An adventurous spirit and a special camaraderie flourished among pilots of all nationalities. They openly admired the achievements of fliers on both sides of the conflict and dropped messages of congratulation, or confirmation of the death of a pilot, as they passed over each other's lines.

EXTRA The Chicago Daily Tribune. FINAL EDITION

THE WORLD'S GREATEST NEWSPAPER

VOLUME LXXVII.—NO. 270. C. MONDAY, NOVEMBER 11, 1918.—22 PAGES. PRICE TWO CENTS.

GREAT WAR ENDS

Washington, D. C., Nov. 11, 3 A. M. (By Associated Press.)—Armistice terms have been signed by Germany, the State department announced at 2:45 o'clock this morning.

The world war will end this morning at 6 o'clock, Washington time, 11 o'clock Paris time. The armistice was signed by the German representatives at midnight.

USE WIRELESS TO GIVE WORD TO SIGN TRUCE

Germany Uses Nearly All of 72 Hours of Grace.

BULLETIN.

Washington, D. C., Nov. 11, 3 a. m.—The momentous news that the armistice had been signed was telephoned to the White House for transmission to the president a few minutes before it was given to the newspaper corre-

OUTLINE OF THE TERMS

(UNOFFICIAL)

Washington, D. C., Nov. 11.—(By the Associated Press.—(The terms of the armistice, it was announced, will not be made public until later. Military men here, however, regard it as certain that they include:

Immediate retirement of the German military forces from France, Belgium, and Alsace-Lorraine.

Disarming and demobilization of the German armies.

Occupation by the allied and American forces of such strategic points in Germany as will make impossible a renewal of hostilities.

Delivery of part of the German high seas fleet and a certain number of submarines to the allied and American naval forces.

REPUBLIC SET UP IN BERLIN BY SOCIALISTS

Manifesto Pledges Government of and for the People.

BERLIN, Saturday, Nov. 9.—(German Wireless to London, Nov. 10, 12:56 p. m.)—By the Associated Press.—The German people's government has been instituted in the greater part of Berlin. The garrison has gone over to the government.

The Workmen's and Soldiers' council has declared a general strike. Troops and machine guns have been placed at the disposal of the council.

Kaiser Flees With Staff to Holland

RED FLAG FLIES OVER ALL BIG GERMAN CITIES

Rebels Continue to Gain; May Exile All Kings.

LONDON, Nov. 10.—With the Social Democratic party leaders and soldiers' and workmen's councils in full control in Berlin, the revolution in Germany is extending rapidly, according to Amsterdam and Copenhagen dispatches.

In most places the desired effect is being achieved without violence or serious disorders.

"Little Bloodshed in Berlin."

An official dispatch received by the Havas agency from Berlin today says:

KRUPP WORKS' CHIEF AND HIS WIFE ARRESTED

Leipsic Joins Revolution.

The Golden Age

As soon as there were planes to fly, daring pilots began wing-walking and looping their way into the hearts of fascinated Americans. Soon, more serious achievements—speed, distance, and altitude records—brought fame to Charles Lindbergh, Amelia Earhart, Wiley Post, Jackie Cochran, Howard Hughes, and others. Handsome airplanes like the Lockheed Vega and the H-1 racer kept upping the standards of performance.

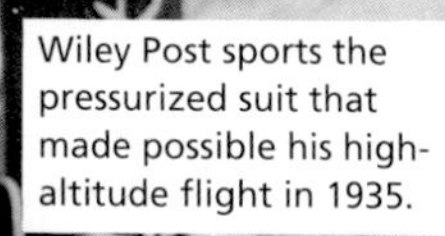

Wiley Post sports the pressurized suit that made possible his high-altitude flight in 1935.

Entertainment in the Air

Stunt Fliers Before World War I

Whirling, spinning, looping—stunt fliers and exhibition teams introduced thousands of people to the airplane after 1910.

Exhibition Teams Two of the earliest exhibition teams were formed by the Wright brothers and Glenn Curtiss to publicize their planes. They flew at country fairs, racetracks, air meets—anywhere that organizers would pay the fee. Other teams, such as the Lincoln Beachey Fliers, were formed by famous aviators.

"Birdmen" (often women) on the teams were either self-taught or graduates of small flying schools. Charles Hamilton, a Curtiss student, was one of the most brazen stunt fliers. His trademark stunt involved ascending to a height of 200 feet, power-diving to within 5 feet of the ground, casually straightening out the plane, and landing in front of the grandstand.

The dangers of daredevil flying—wing walking, hanging from airborne planes, changing planes in midair—led to many short careers. Of the five original fliers on the Wrights' exhibition team, only one, Frank Coffyn, survived the two-year contract.

Women Take Wing In September 1910, Glenn Curtiss placed a throttle block on a plane to prevent a woman student from leaving the ground. But future barnstormer Blanche Scott managed to get 40 feet into the air despite this handicap. Curtiss soon accepted her as a member of his exhibition team.

Two other stunt pilots, Ruth Law and Katherine Stinson, faced prejudice regarding their ability to control a plane by setting records and matching the men's fate-tempting feats. Law was the first woman to fly at night and the first woman to loop the loop. In 1917, she set a women's altitude record of 14,700 feet—though she wanted to compete with men as an equal rather than in a separate category.

Katherine Stinson came from a flying family. Her sister Marjorie flew, as did their brother Eddie, who in 1916 figured out how to recover a plane from a spin. Before that, spins during stunts killed many pilots. Katherine Stinson developed her own stunt, called a dippy twist loop. At the top of a loop, she rolled her aircraft wing over wing. In 1917, Stinson flew 610 miles to break the nonstop distance record.

Former drama critic Harriet Quimby became the first licensed woman pilot in the United States. Nicknamed "the Dresden-China Aviatrix," she was also the first woman to fly the English Channel. She died when turbulence toppled her out of her plane during the Boston-Harvard aviation meet in 1912.

In 1912 Harriet Quimby, also a stunt flier, became the first woman to fly the English Channel. Though an Englishman offered to make the flight for her (disguised in Quimby's signature purple silk costume), she preferred doing it herself.

The *Vin Fiz* Stunt fliers like Calbraith Rodgers set many altitude and distance records. In pursuit of a $50,000 prize for the first cross-continent flight made within 30 days, offered by William Randolph Hearst, he left New York on September 17, 1911. His plane was named the *Vin Fiz,* a grape drink sold by his sponsor, the Armour Company.

Rodgers was followed by a train that carried a mechanic, spare parts, and his wife, Mabel. Rodgers crashed 19 times and made 69 stops due to mechanical trouble before he landed the *Vin Fiz* in Pasadena. He had one leg in a cast, only a rudder and a single wing strut left from his original Wright plane, and had taken 49 days to cross the country—thereby forfeiting the prize. Still, a crowd of 20,000 cheered his arrival.

Perhaps the most patched-up plane ever, the *Vin Fiz* crashed 19 times on its cross-continent trip in 1911. Despite the plane's radiator leaks, burst engine cylinders, and a plunge into a chicken coop, pilot Calbraith Rodgers finished the trip, grin and cigar intact, in 49 days.

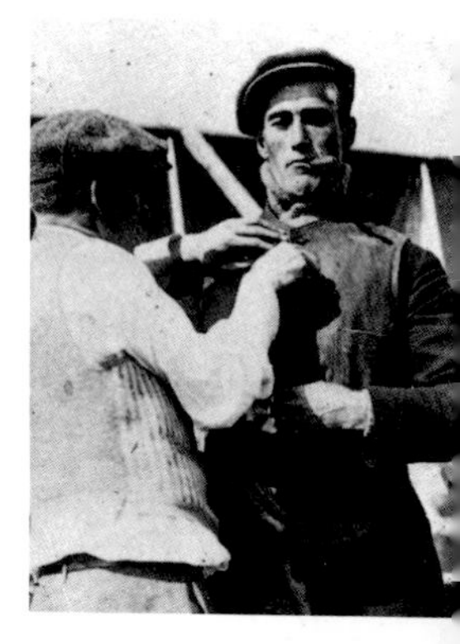

The Barnstormers

Lincoln Beachey, Ralph Johnstone, Harriet Quimby—the list of stunt pilots killed before World War I was long. After the war, demobilized pilots joined the stunt fliers. Together they were called barnstormers (the name evokes flying in cow pastures and sleeping in barns). Surviving on shoestring budgets, these aviators flew anytime, anywhere, charging from $2 to $25 for a ride.

Curtiss "Jennies" Barnstormers often purchased war-surplus Curtiss JN-4Ds, priced at $5,000 during the war but available for about $600 after 1918. Airplane maintenance was improvised: a broken wing spar might be spliced with a piece of pine from an orange crate, and a worn engine arm might be replaced with the steel barrel of a fountain pen.

An acrobat hangs by his legs from the crossbar of the landing gear of a British Avro 504k. The Avro 504 series of two-seater trainer aircraft were first flown in 1913, and variants of the 504k saw service in both world wars.

Wing Walking and More Stunt fliers found inventive ways to defy death. Ormer Locklear hung upside down from the plane's lower wing, his legs hooked over the wing's skid. Phoebe Fairgrave made a record jump (with a parachute) from 5,000 feet in July 1921. Other stunts included transferring from plane to plane in midair and flying upside down.

Often the barnstormers wandered into their best shows. Charles Lindbergh got lost in Mississippi on his way to Texas. Landing in a rough field, he damaged his plane's propeller. Local residents helped him fix it, and he started taking them on rides. The locals clamored for more, so he stayed for two weeks, flying stunts and joy-riding with passengers. He turned down only one request: a woman wanted to know how much he would charge to take her to heaven and leave her there.

Air Races: Ushering in the Golden Age

Ever since the Rheims air meet of 1909, aviators had raced each other for glory and prize money. Newspapers and wealthy private citizens posted most of the prizes, which ran as high as $25,000. Some of the races were based at a single location, such as the King's Cup in the British Isles or the National Air Races in the United States. Others were international, like the Schneider Trophy.

Even the magician Houdini performed stunts in his Voisin plane.

The Schneider Trophy

Established in 1913 by Jacques Schneider of France, the Schneider Trophy promoted the development of high-speed seaplanes, which at the time formed the backbone of air transport craft in Europe. Each year, the prize was given to the fastest plane covering 350 kilometers over a triangular course. Government sponsorship funded the development of these craft (sometimes called *hydroaeroplanes*).

Governments eventually balked at the cost, but wealthy patriots stepped in. In 1930 the British government decided not to finance the design and construction of a new craft for the Schneider race; an indignant Lady Lucy Houston donated £100,000 and let fly with her opinions: "I am utterly weary of the lie-down-and-kick-me attitude of the Socialist Government . . . I live for England and I want to see England always on top."

Her investment and her gutsiness were rewarded. In 1931, Britain won the Schneider Trophy with a Supermarine S.6B using a Rolls-Royce R (racing) engine, which delivered 2,350 horsepower. It was Britain's third win in three years and thus the last Schneider awarded—any country winning three times in five successive attempts was to retain the trophy in perpetuity.

The world records in air speed attained during the Schneider races highlight the swift pace of technical improvements. In 1913, for example, France won the trophy with an average speed of 46 miles per hour. In 1931 Great Britain won with an average speed of 340 miles per hour.

The Pulitzer Race

Another famous race of the 1920s was the Pulitzer, established by the Pulitzer brothers of St. Louis. The first Pulitzer was won on November 25, 1920, by Captain Corliss Moseley of the U.S. Air Service. Like the Schneider, the Pulitzer competition was dominated by pilots and planes from the armed forces, and was flown over a triangular course.

The 1922 Pulitzer marked the debut of the Curtiss R-6 racing biplane. U.S. Army pilots won both first and second place flying this aircraft. Another Curtiss racer, the R2C-1, appeared in the Pulitzer of 1923. Once again, the U.S. armed forces took first and second place in the new plane.

Speed was the goal at the National Air Races. Founded in the 1920s, the races included a closed-course circuit for the Thompson Trophy and a cross-country competition for the Bendix Trophy.

Flying Boats and Seaplanes

Flying boats and seaplanes garnered world records in speed and made pioneering flights during the golden age. There is a distinction between the two craft: a seaplane has floats for landing on water; a flying boat's fuselage can float.

A Curtiss NC flying boat was the first heavier-than-air craft to fly the Atlantic. In May 1919, three of these "Nancies" left Newfoundland for the Azores. NC-4, the only one to complete the trip, arrived in 15 hours. It continued on to Portugal and England.

In the 1920s, several seaplanes set speed records in the Schneider Trophy race. The last of these, the Supermarine S.6B, was the forerunner of the successful Spitfire fighter of World War II.

Clipper-design flying boats such as the Martin M-130 and the Boeing 314 made some of the earliest transpacific flights in the 1930s. The era of flying boats and seaplanes faded with the coming of long-range, four-engine airplanes in the early 1940s.

The Curtiss R3C-2, piloted by James Doolittle, won the 1925 Schneider Trophy with a speed of 232 miles per hour. One day after the race, Doolittle flew it to a world speed record of 245 miles per hour.

The Women's Air Derby

This lineup of participants in the Women's Air Derby of 1929 includes winner, Louise Thaden (far left), and Amelia Earhart (fourth from right).

Women pilots perceived themselves as emissaries of aviation and liberators of their gender. Pilots such as Amelia Earhart decried the separate endurance, speed, and height records maintained for women. They were skilled and courageous aviators, not "sweethearts of the air" or "petticoat pilots." They wanted the opportunity to race men.

It was a women's cross-country event, the 1929 Women's Air Derby, that forced the aviation industry to accept women as serious pilots. Initially, the all-male organizing committee suggested that the women start in Omaha rather than Santa Monica in order to be spared the danger of crossing the Rocky Mountains. Earhart and the other 19 contestants rebelled, refusing to fly unless the course remained as originally charted. The organizers reconsidered, and Louise Thaden won the race.

The Bendix Races

Their success in the Bendix race showcased the skills of women pilots. The Bendix was a transcontinental speed dash dominated throughout the 1930s by civilian aircraft. With $15,000 in prize money, top competitors always entered.

In 1936, Louise Thaden won first place in the Bendix with a time of 14 hours, 55 minutes, between Los Angeles and Cleveland. She flew a Beech "Staggerwing" to beat the racing and twin-engined aircraft flown by the men.

A woman won again in 1938, when Jacqueline Cochran took the race in a single-seater Seversky pursuit plane. After a wad of paper lodged in the plane's fuel pipe, she flew the plane with one wing tipped higher than the other to ensure an adequate fuel supply.

Even that achievement, however, was questioned by skeptics. She emerged from the plane "looking so fresh" that it was rumored she had picked up a male pilot, who flew the race while she slept, and then had dropped him off before reaching the finish line. How she could have landed twice and still have beaten nine men was a mystery the rumor did not address.

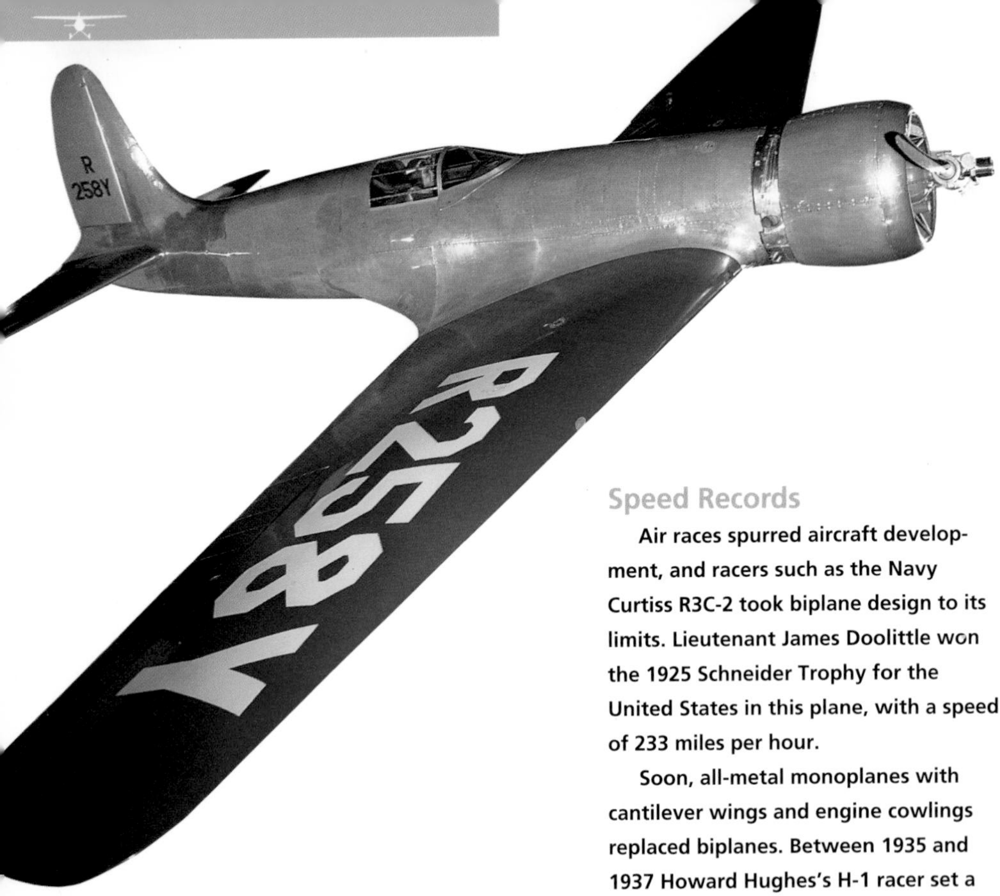

Speed Records

Air races spurred aircraft development, and racers such as the Navy Curtiss R3C-2 took biplane design to its limits. Lieutenant James Doolittle won the 1925 Schneider Trophy for the United States in this plane, with a speed of 233 miles per hour.

Soon, all-metal monoplanes with cantilever wings and engine cowlings replaced biplanes. Between 1935 and 1937 Howard Hughes's H-1 racer set a world's landplane speed record of 352 miles per hour and set a new transcontinental record of 7.5 hours. Only two years later in 1939, a modified German Messerschmitt fighter was clocked at 469 miles per hour.

Several features of the Hughes H-1 racer—close-fitting engine cowling, retractable landing gear, sleek aluminum skin, and smoothly joined wings and fuselage—were designed to reduce drag and became common features of later plane designs.

Distance Records

Throughout the 1920s, aviators conquered one distance after another. The first solo flight between Great Britain and Australia, the first transpacific flight, the first crossing of the Tasman Sea—the list goes on and on.

The speed with which airplanes crossed these geographical boundaries made the world seem very small. No country, however remote, could remain isolated with airplanes winging across seas and mountain ranges. Through the aviators' eyes, the public could experience exotic, distant places.

Macready and Kelly's Transcontinental Flight

The first transcontinental crossing in a single day was accomplished in 1923 by Lieutenants J. A. Macready and O. G. Kelly of the U.S. Air Service in a Fokker T-2 aircraft, a partially fabric-covered plane with plywood-veneered wings. At one point, Kelly disassembled a faulty voltage regulator in midflight while Macready controlled the plane from the back cockpit—a dangerous operation.

Throughout the 27 hours the pilots were in the air, people along the route listened for the plane and peered eagerly into the sky. In San Diego, throngs gathered to watch them land. Car horns honked, and factory whistles blew.

Anthony Fokker congratulated them with these words: "Your flight is a milestone in the development of commercial aviation. In ten years the route you flew will be covered by aerial passengers and freight service just as Blériot's route across the English Channel is today."

Commander Richard Byrd outfitted himself in heavy furs and thick boots for his polar flights.

Byrd's Polar Flights

In 1926, Commander Richard Byrd of the U.S. Navy flew over the North Pole with his pilot, Floyd Bennett. In their Fokker F.VII-3m they carried a handmade sledge and rubber boat as emergency gear.

In his account of the flight, Byrd spoke of the difficulty of navigating near the pole. No landmarks appear on the icefields; the pair had to rely on the sun, stars, and moon to calculate their position. Their chief concern was to fly due north—a goal that an ordinary magnetic compass could not help them achieve. Such a compass would have pointed toward the north magnetic pole, which lies more than 1,000 miles from the geographic pole.

Byrd and Bennett relied on a sun compass, essentially a reversal of a sundial: the shadow of a vertical pin mounted on the tip of the hand of a 24-hour clock, indicated the plane's direction. Byrd knew that he could not have reached the pole without it.

Next, Byrd headed for Antarctica in November 1929; this time his pilot was Bernt Balchen. Flying the 7-ton Ford Tri-motor over a ridge of mountains and up to the polar plateau was the most dangerous part of the trip; the team had to drop food overboard to lighten the plane enough to clear the mountains. Despite these difficulties, the trip was a success.

This Fokker F.VII-3m tri-motor monoplane, the *Josephine Ford*, flew Byrd and Bennett to the North Pole. Only three days later, American explorer Lincoln Ellsworth completed a similar flight at the controls of the airship *Norge*.

Earhart's Around-the-World Flight

Earhart's Atlantic Crossing

Thompson Trophy

Nonstop Transoceanic Flights

In 1913 Lord Northcliffe and the London *Daily Mail* offered £10,000 for the first nonstop aerial crossing of the Atlantic in either direction. The trip was to be completed in 72 hours, with takeoff and landing points anywhere in the United States or Canada and the British Isles. The prize remained uncaptured throughout World War I.

In 1918 Lord Northcliffe renewed the prize and generated a spate of contenders. A takeoff from Newfoundland provided the shortest route. By the spring of 1919, freighters were hauling crated planes from Britain to St. Johns, Newfoundland; aviation teams were scouting the best meadows to level into landing fields; and pilots were anxiously watching one another's preparations.

The first attempts ended in crashes, none fatal. Then a British team, Captain John Alcock and Lieutenant Arthur Whitten-Brown, took off from Newfoundland on June 14 in a converted Vickers Vimy bomber, which had a cruising speed of 90 miles per hour and a range of 2,400 miles.

The flight, though successful in winning the prize, was not without incident. When the Vimy entered a bank of thick fog, Alcock nosed the plane up to 12,000 feet to regain a view of the stars. Feeling as if he were flying "in a bottle of milk," Alcock became disoriented. Plunging through the clouds, the Vimy broke into the clear and leveled out at just 50 feet above the ocean. Sixteen hours after takeoff the team crash-landed in a bog in Ireland and won the *Daily Mail* prize.

Bendix Trophy

Auditorium Hotel Trophy

The glory days of aviation sparked lively competition among airplane manufacturers and aviators. Shown above are some of the prestigious trophies offered by a wide range of sponsors in both the public and private arenas. The "around the world flight" trophy on the far left—intended for Amelia Earhart—remains unclaimed.

Charles Lindbergh

Another cash award—the Orteig Prize of $25,000 for the first nonstop flight between Paris and New York—was the catalyst for the epic flight of Charles Lindbergh.

Planning the Flight Lindbergh was a little-known airmail pilot with a tight budget. Even after he found promoters for his effort, none of the large aircraft companies, such as Fokker and Bellanca, would commit their reputations to his hands by building a plane for him. Finally he heard about a high-wing monoplane built by a company called Ryan near San Diego. They agreed to build a special version of this aircraft for Lindbergh.

With the plane and backers secured, Lindbergh focused on planning the flight. The key issue was weight. His greatest hedge against failure was extra fuel; he had decided against bringing a navigator, who would take up some of the weight allowance. But the more fuel he carried, the riskier the takeoff in a heavily loaded plane. It was a risk he decided to take.

Then there was the question of navigation. As an airmail pilot, Lindbergh had always relied on lakes, towns, bends in rivers, and railroad tracks to find his way. He knew it would be different flying the Atlantic, where he would have to navigate by timing and calculation rather than by landmarks. Still, he wasn't that worried. Europe was a continent; he figured he couldn't miss it.

The *Spirit of St. Louis*

The airplane Lindbergh piloted on his flight from New York to Paris was a standard Ryan model customized for the trip. The fuselage and wings were lengthened and reinforced to carry a 425-gallon fuel load. The cockpit was located far back in the fuselage (away from the forward fuel tanks for safety). To see forward, Lindbergh used a specially designed periscope or turned the plane slightly to look through a side window.

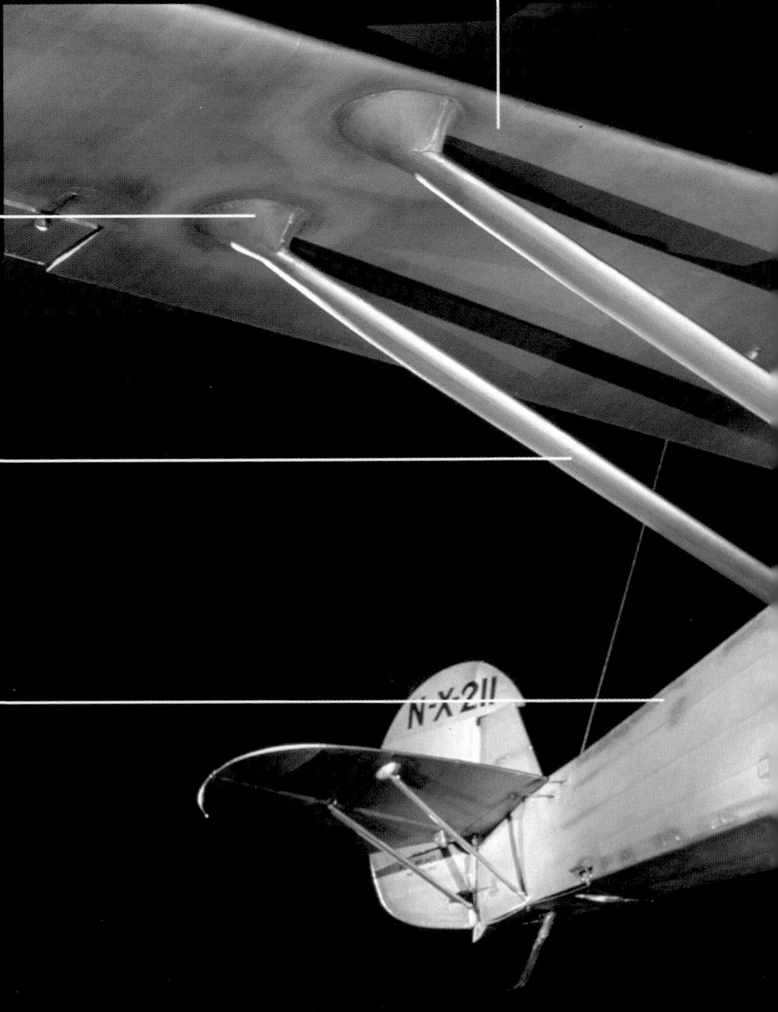

The fuel tanks are mounted in the wings and behind the engine in the fuselage.

Streamlined surfaces were designed to reduce drag.

The struts brace high, spruce-framed wings.

The fuselage is constructed of steel tubing. Taut cotton fabric covers the fuselage and the wings.

Lindbergh sat amidships, behind the engine and the main fuel tank. The side windows and a retractable periscope are the only ways to see out of the cockpit.
The aluminum engine cowling is decorated with flags of the countries Lindbergh visited in this airplane.
The air-cooled Wright Whirlwind engine, capable of 223 horsepower, was selected by Lindbergh for its reliability.
Spirit of St. Louis

Cockpit of the *Spirit of St. Louis*

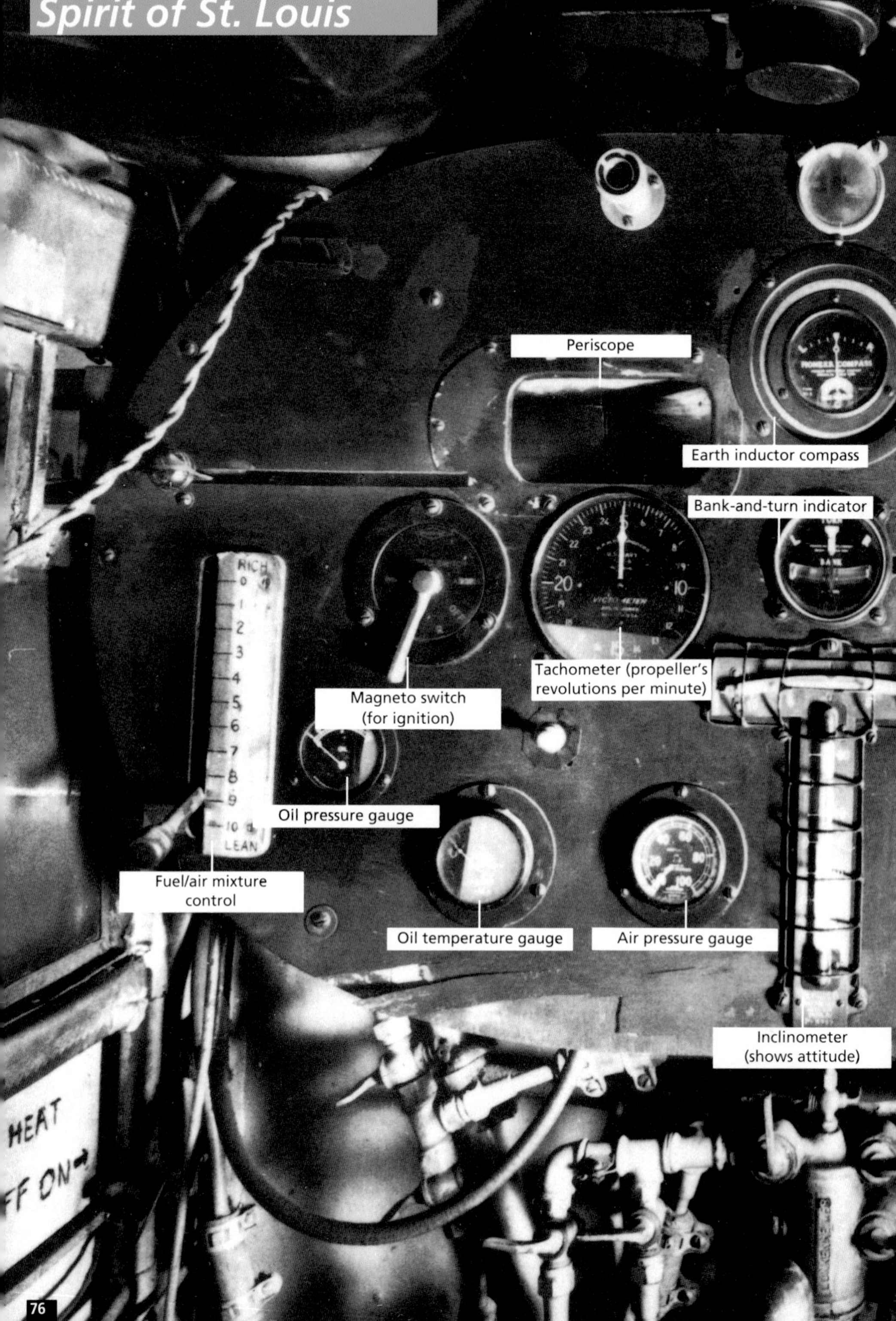

Because the ocean shows no landmarks, Charles Lindbergh relied entirely on these instruments to navigate from New York to Paris.

Lindbergh's flight inspired a variety of memorabilia, from *Spirit of St. Louis* lamps and clocks to the badges shown below.

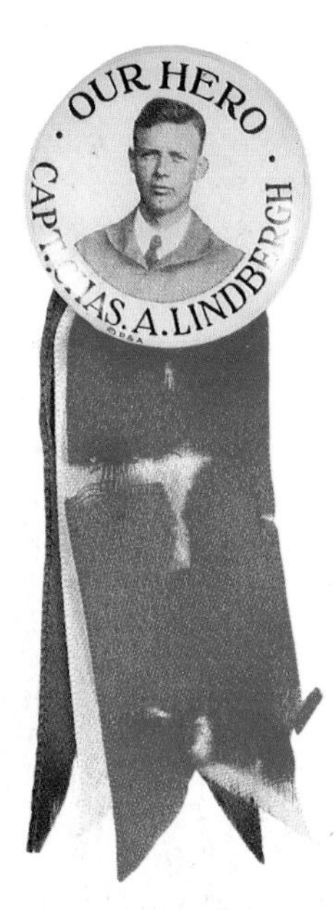

The Competition The field of Orteig Prize competitors, including many star pilots from World War I, was narrowed by disaster. René Fonck, a French ace trying for the prize, crashed when the landing gear on his Sikorsky collapsed (carrying the weight of a bed, four men, and a celebration dinner). Richard Byrd's Fokker crashed on a test flight (without casualties). Then, on April 26, 1927, another team was killed in a practice takeoff with a full load.

Lindbergh and the other contenders carefully watched for news of a French team led by World War I ace Charles Nungesser, attacking the challenge from east to west. On May 8, 1927, Nungesser took off from France. Rumored sightings were reported all the way to the United States, but he never arrived. The last confirmed view of his airplane, *L'Oiseau Blanc,* was by a plane near the European coast.

The Flight: May 20, 1927 Lindbergh took off in bad weather while the search for Nungesser continued. He struggled to get his plane off the muddy ground: "The *Spirit of St. Louis* feels more like an overloaded truck than an airplane," he wrote in his log. Nearing Nova Scotia, he decided to turn back if he encountered fog there and could not check his course by landmarks. But the air was clear, and he headed for Europe.

His biggest challenge was staying awake for more than 33 hours: "My eyes feel dry and hard as stones. The lids pull down with pounds of weight against the muscles." Fog and clouds surrounded him: "Everything is uniform blackness, except for the exhaust's flash on passing mist and the glowing dials in my cockpit, so different from all other lights . . . My world and my life are compressed within these fabric walls."

When dawn came, Lindbergh sighted fishing boats and circled one, vainly calling, "Which way is Ireland?" He reached the Irish coast only 3 miles off course, then proceeded to Paris. His first thought when he saw the landing field was typical of an airmail pilot: "It's a shame to land with the night so clear and so much fuel in my tanks."

So many people rushed the plane as he landed that he had to turn off the propeller to keep from harming them. After being feted in Europe, Lindbergh returned home in a navy ship sent by President Coolidge.

In the United States, Lindbergh found himself an overnight sensation. Nearly four million people lined the streets in Manhattan for his ticker-tape parade. Articulate, dedicated, an honest pioneer, Lindbergh was admired for his character as well as his flight.

Charles Lindbergh

After his transatlantic flight, Charles Lindbergh became an instant hero and put his fame to work by promoting support for the development of aviation.

At the request of the Guggenheim Fund for the Promotion of Aeronautics, Lindbergh toured the United States in 1927, flying the *Spirit of St. Louis*. For three months he visited cities and small towns and dropped messages over the towns that he couldn't visit.

In December 1927 Lindbergh agreed to fly as a U.S. goodwill ambassador to several Latin American countries. While on this tour he met Anne Morrow, whom he married in 1929. She became his partner in flight and accompanied him on several international trips to chart new routes for airlines. She acted as copilot and radio operator. A gifted writer, she published several books.

On March 1, 1932, a nightmare began for the Lindberghs when they discovered that their 20-month-old son had been kidnapped. Eventually the body of the child was found, and a murder suspect was convicted and executed. The Lindberghs—exhausted from constant publicity—left for Europe to regain a private life.

During the first years of World War II, Lindbergh was harshly criticized for supporting noninvolvement. But by 1944 he was flying combat missions as a technical adviser in the Pacific Theater. After the war, Lindbergh withdrew from public attention, serving as a consultant to various airlines and as a spokesperson for nature conservation until his death in 1974.

Nicknamed "Lady Lindy" after making her first transatlantic flight as a passenger, Amelia Earhart wanted to prove her skill by flying the Atlantic on her own.

Amelia Earhart

One year later, Amelia Earhart also flew the Atlantic—but as a passenger. A social worker and pilot, she took the place of Amy Guest, heir to a Pittsburgh steel fortune, whose family had forbidden her to make the trip. Guest agreed to give her spot to another woman. Earhart jumped at the opportunity.

At the time, only six people had flown the North Atlantic in airplanes since Alcock and Whitten-Brown in 1919. In 1927 alone, 19 people had died attempting it. The crossing proved to be uncomfortable and cold; Earhart had to crouch at the navigation table behind extra fuel tanks. At dawn, with a dead radio, she and the pilot, Bill Stultz, bombarded a ship with messages weighted with oranges. They asked, in vain, for the sailors to paint a message on the deck detailing the plane's position. By luck, they found Wales and then commenced a two-week round of celebratory teas, speeches, visits to the theater, exhibition tennis matches, and Parliament.

Earhart did not feel like a heroine; she equated her usefulness on the flight with that of "a sack of potatoes." Yet she took seriously her new role as a spokeswoman for aviation, writing articles for *Cosmopolitan* such as "Should you let your daughter fly?" And she took every opportunity to gain flight experience.

In September 1928 Earhart flew to Los Angeles to visit her father and then flew back to New York. It was, again, a record: the first solo round trip across the continent. The lack of navigational aids made it difficult to distinguish one small town from another at 100 miles per hour. When her map blew out of the

cockpit, Earhart caused a stir by landing on the main street of a small town to get another one.

Like Lindbergh, Earhart served as a public relations figure for the Transcontinental Air Transport company in 1929. She also bought a second-hand Lockheed Vega, though she soon replaced the "clunk" with a new one. After participating in the Women's Air Derby at the end of the year, Earhart kept striving toward new goals. On June 25, 1930, she set three women's world speed records—over a 3-kilometer course, over a 100-kilometer course, and with a payload.

In a different Lockheed Vega, Earhart completed the first solo flight from Hawaii to the U.S. mainland in 1935.

On May 21, 1932, Amelia Earhart landed in Ireland in this Lockheed Vega, the first woman to fly solo across the Atlantic. In the same year and in the same plane, she made the first nonstop solo transcontinental flight by a woman, from Los Angeles to Newark.

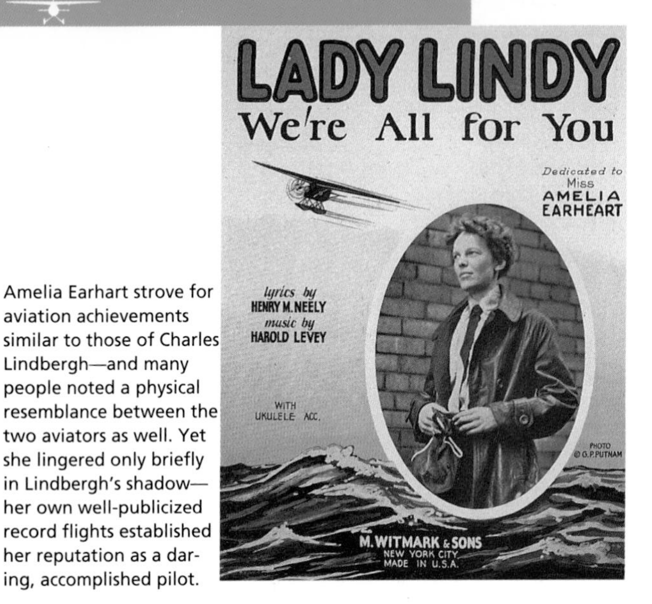

Amelia Earhart strove for aviation achievements similar to those of Charles Lindbergh—and many people noted a physical resemblance between the two aviators as well. Yet she lingered only briefly in Lindbergh's shadow—her own well-publicized record flights established her reputation as a daring, accomplished pilot.

Earhart's Transatlantic Flight Though Earhart had proven her flying ability again and again, she coveted another transatlantic flight—this time as pilot. After her new Lockheed Vega was modified for the flight, she took off on May 20, 1932, from Newfoundland. The weather was rough: she endured a violent thunder storm, wing icing, and, at one point, a vertical drop of nearly 3,000 feet. Engine vibrations forced her to land in Northern Ireland rather than Paris. Still, her 15-hour flight set three records: the fastest crossing of the Atlantic, the first transatlantic flight piloted by a woman, and the first solo crossing by a woman.

Earhart was showered with honors. She met the pope, Mussolini, and the king of Belgium. She also was given the National Geographic Society's gold medal, the first time it was awarded to a woman.

Hegenberger and Maitland's Pacific Flight

Reaching Hawaii (not yet the fiftieth state) by air was another coveted achievement. In June 1927, Lieutenants Albert Hegenberger and Lester Maitland flew a Fokker C-2 trimotor from Oakland, California, to Honolulu, Hawaii, in 25 hours, 50 minutes.

Amelia Earhart flew the reverse path in mid-January 1935—the first solo flight from Hawaii to the mainland. Her Lockheed Vega made the flight in 18 hours.

Within 12 months of Lindbergh's flight, applications for U.S. pilot licenses tripled, and the number of U.S. airline passengers quadrupled.

Around-the-World Flights

The Douglas World Cruisers

After World War I the U.S. government was ambivalent about the potential of air power and left the issue on a back burner. Colonel Billy Mitchell, an ardent promoter of the U.S. Air Service, countered by arranging displays of aviation prowess. Army pilots usually won the speed races in the early 1920s, but by 1923 Mitchell had set his sights on a new publicity stunt, an around-the-world flight—a feat attempted by many nations but not yet accomplished.

Mitchell planned to form a team of four planes powered by Liberty engines. The first step was to plan the route and send an advance team to find landing fields and stash crates of spare engines, parts, and tools at each stop. The Liberty engine was reliable for only 50 hours of flight, which meant that a dozen complete engine changes would have to be made for each plane over the course of the flight.

Mitchell chose a small California airframe builder, Donald Douglas, to build the planes. It was a huge coup for Douglas—four years earlier he had opened his aircraft business in the back of a barber shop with $600 in capital.

The Douglas World Cruiser *Chicago* (below), together with the *New Orleans,* successfully circumnavigated the globe in 1924—it took almost six months to finish the journey. The planes had no radios and only minimal navigational equipment but were assisted by the U.S. Navy throughout the trip.

The finished World Cruisers had a wingspan of 50 feet, a ceiling of 8,000 feet, and a normal cruising speed of 90 miles per hour. They were open-cockpit biplanes with fittings for either floats or wheels, depending on the location. They had no radios, navigational aids, or weather-forecasting equipment. Their instruments consisted of a compass, an altimeter, and a bank-and-turn indicator.

The *Seattle*, the *Chicago*, the *Boston*, and the *New Orleans* left Seattle on April 6, 1924, on their way to Alaska. The *Seattle*'s pilot, blinded by snow, flew into the ground, damaging the airplane beyond repair. In the Gulf of Tonkin, the *Chicago*'s engine blew up; the pilot hired a fleet of *sampans* (Chinese skiffs) to tow him to the dock, where a new engine was delivered by a navy destroyer. Later, the *Boston* was forced down in the North Atlantic. The *Chicago* and the *New Orleans* reached Seattle on September 28, 1924, after flying 26,345 miles in six months. Each plane had used nine engines and had relied heavily on the navy's support.

The *Question Mark*

Lowell Smith was one of the best cross-country pilots in the U.S. Air Service. In 1919, he successfully completed a transcontinental air race. Later, he established an endurance record of 37 hours in a de Havilland aircraft.

In 1929, Smith, flying the *Question Mark,* set an endurance record of 151 hours in the air. The record was dependent on midair refueling, and Smith had been the first pilot to risk this procedure in 1923. He used a 40-foot, metal-lined, flexible steam hose for the purpose. One false move—a few drops of gas on the hot engine—and the flight would have been over.

Wiley Post's Record Flights

On June 23, 1931, Wiley Post set out from New York in his Lockheed Vega, the *Winnie Mae,* to circle the world. He and his navigator passed over Newfoundland, England, Germany, Russia, Siberia, Alaska, and Canada, before landing in New York 8 days and 15 hours after they had set out. The flight amazed the press, and so did Post, a former wing-walker with an eyepatch and a devilish grin.

Post flew the *Winnie Mae* around the world again, two years later—without a navigator. With the aid of a radio compass and an autopilot, he made the flight in 7 days, 19 hours.

For his next challenge, Post modified the *Winnie Mae* for long-distance, high-altitude flight. Because the Vega could not be pressurized, Post asked the B. F. Goodrich Company to develop a pressurized suit for him. By equipping the plane with an engine supercharger, Post hoped to cruise at high altitude in the jet stream. On March 15, 1935, he flew from California to Cleveland in 7 hours, at an average speed of 277 miles per hour.

Wiley Post's modified Lockheed Vega, the *Winnie Mae*, set two records for around-the-world flights, in 1931 and 1933. With Post again at the controls, the airplane made a historic high-altitude flight from California to Cleveland, speeded by the jet stream, in 1935. Post wore a three-layer pressurized suit with an oxygen-equipped helmet.

Amelia Earhart's Last Flight

Earhart's announcement of a world flight, in 1937, seemed a portent of disaster. Old friends claimed she was caught up in the "hero stream" of fliers, each performing bigger and braver feats. She was under pressure to equal them.

Amelia Earhart decided to make the most dangerous flight of her life in her first two-engine plane, a powerful Lockheed Electra. She had little experience with the plane but refused to delay the flight. To make matters worse, her navigator, Fred Noonan, was known for his fondness for alcohol.

The first leg of the trip, from California to Hawaii on March 17, 1937, was completed successfully. But during the takeoff from Honolulu, the Electra crashed on the runway, resulting in $25,000 of damage. Earhart set off again on June 1, but in the other direction: during the delay, weather conditions worldwide had changed, making the trip more feasible with the longer of the two Pacific crossings moved to the trip's end. In the first week, she flew across the Atlantic to Africa. By June 18, she took off from Calcutta, with monsoon rains beating hard against the Electra's wings.

After reaching Lae, New Guinea, Earhart wrote an article about the journey. In 40 days she had flown 22,000 miles, crossed the equator 3 times, and landed 22 times, a pace she described as "leisurely."

On July 2, she and Noonan left Lae bound for Howland Island in the Pacific. For seven hours radio operators in Lae were in touch with the pilot, but navy ships could never establish communication. Earhart had removed or lost the radio equipment needed to communicate with them. She was close: the cutter *Itasca* could hear her loud and clear, but she could not hear their frantic messages to her. After stating that gas was running low and that she was unable to find the island, Earhart slipped into silence.

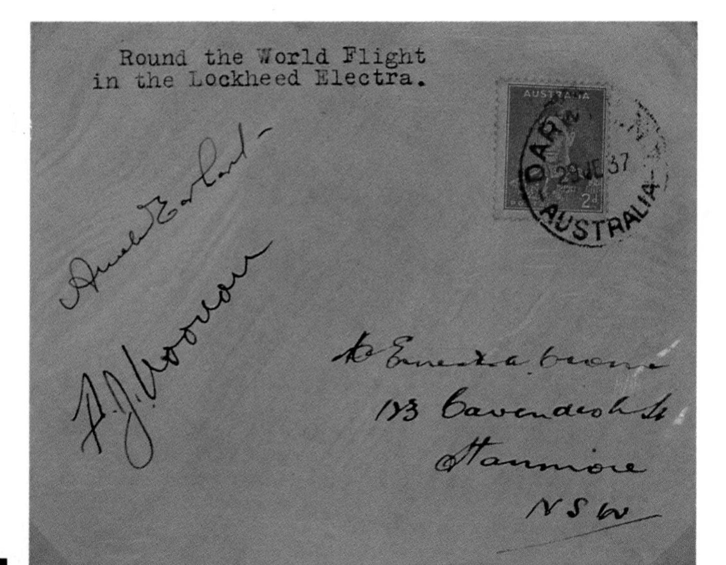

Earhart sent souvenirs of her most ambitious flight to friends all over the world.

The Fascination with Earhart

How could Amelia Earhart, the premier female pilot, disappear in the most expensive and best-equipped privately owned aircraft of the time? The answer may lie in preparations for the flight.

Perhaps the worst miscalculation was Earhart's cavalier attitude toward radio equipment. She never took the time to learn Morse code or to understand the Electra's radio system, although several experts encouraged her to do so.

When her plane disappeared, a massive rescue effort was mounted to find the lost aviator. Ten ships and 65 airplanes combed 250,000 square miles of the Pacific in a 16-day search. The legend of Earhart had begun.

In the story line of a 1943 propaganda film, a famous pilot volunteers to become "lost" in the Pacific in order to spy on the Japanese. Rumors spread that Earhart had been on such a mission for President Franklin D. Roosevelt. Or perhaps she had been executed by the Japanese or imprisoned in Tokyo until the end of the war, when she was shipped home incognito. Others suggested that her disappearance was a publicity stunt by her husband, publisher George Putnam.

As late as 1993, a World War II veteran claimed to have seen her plane in the jungle, setting off yet another string of searches. The very real possibility that Earhart may have lost her way and run out of fuel lacks the power of myth.

Air Transport

After World War I, the aviation business slowly began to flourish. Though the U.S. airmail system made a limping start in 1918, it eventually became a highly profitable enterprise for the early airlines. The feats of famous aviators—and more dependable, comfortable airplanes like the classic DC-3—inspired a boom in passenger transport. The airplane also proved its usefulness in dozens of new businesses.

SEE

AMERICA BY AIR

SUGGESTED AIRPLANE TOURS of the UNITED STATES and MEXICO

Designed to lure new passengers, this travel brochure promotes twelve pleasure trips on five different airlines.

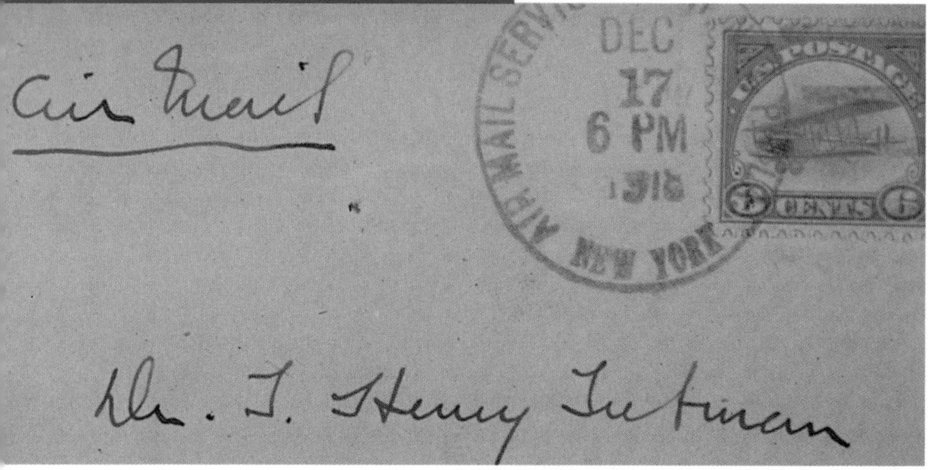

Before passenger service became a money-making enterprise, airmail made profits for the early U.S. airlines. Now a taken-for-granted convenience, the original airmail system required vast planning efforts and a nervy set of pilots to make it work.

Airmail

The Birth of Airmail

On May 15, 1918, with President Woodrow Wilson and thousands of others watching, the first airmail flight in U.S. history began—a bit awkwardly. First, the airplane, a Curtiss Jenny, belched blue smoke, sputtered, and would not start—until someone noticed it was out of fuel. Gas tanks filled, the pilot took off, and 140 pounds of airmail was finally on its way from Washington, D.C. to New York. Then, as the crowd watched in disbelief, the plane headed off in the wrong direction.

The pilot later landed in a plowed field in Maryland, completely lost, and the first airmail had to be sent to its destination by train.

Army pilots soon improved on this inauspicious beginning and by August, when the U.S. Post Office took over the service, airmail was becoming well established in the eastern part of the country. In the first year, 1,208 flights were completed.

Airmail Planes To build up its fleet, the Post Office obtained surplus army planes, such as the de Havilland D.H.4, at almost no cost. By 1921, half of the 98 airmail planes in service were D.H.4s. Though they had to be renovated to carry mail, and were far from ideal, they were the best option financially.

Two other workhorses were the Douglas M-2, introduced in 1926, and the Pitcairn Mailwing, which entered service early in 1928. The M-2 could carry up to 1,000 pounds of cargo. The Mailwing, a trim, open-cockpit biplane with a lightweight frame and a reliable Wright Whirlwind engine, also gained a reputation as an efficient cargo carrier.

Airmail Spans the Country By the time the service was two years old, airmail reached from the east coast to Omaha, Nebraska. Technically, though, the service was not entirely delivered by air. At night, the airplanes landed and the mail was "entrained." Without flashing beacons or lighted fields, the pilots could not navigate after dark. Trains were needed to keep the mail moving.

This situation changed in February 1921, following the election of President Warren G. Harding. Harding was an outspoken opponent of airmail service. The U.S. Post Office needed a dramatic breakthrough to gain the attention of Congress and ignite public support for airmail. The bold plan was to fly mail coast to coast by plane, without using trains at night.

The Post Office arranged for two airplanes to take off from Long Island, heading west, on February 22, while two from San Francisco flew east. Relays of planes would carry the mail in each direction toward a meeting point in Omaha.

One eastbound plane crashed, killing the pilot. The other landed successfully in North Platte, Nebraska, where pilot Jack Knight waited to carry the mail into Omaha. Following railroad tracks and flickering bonfires along his route, Knight reached Omaha at 1:10 A.M. on February 23—only to find that his journey was just beginning.

The westbound plane was stuck in a snowstorm in Chicago. The airmail chain across the country was broken. Unless something was done, the route would have to be completed by train.

A de Havilland light bomber manufactured under license in the United States during World War I, the D.H.4 was retooled for use as an airmail plane after the war. A compartment to hold 500 pounds of mail was built into the front seat, and the pilot moved to the back.

It took courageous pilots to pioneer the first air-mail routes. Of the first 40 pilots hired by the Post Office in 1918, only 9 were still alive in 1925.

So Knight, who had never flown east of Omaha, climbed back into his plane. Flying low, to keep an eye on the ground through the thick snow, Knight made it to Iowa City. At 6:30 A.M., with a full tank of gas, he took off for Chicago, landing his D.H.4 there at 8:40 A.M. A pilot in Chicago flew the mail the last leg to New York.

Seven pilots had covered 2,660 miles through bad weather—Jack Knight flying 710 of those miles himself. San Francisco mail had reached New York in 33 hours, instead of the 108 hours formerly required using trains at night. Headlines exploded with the story, and President Harding lent his support to the U.S. Post Office's request for money to light the airway.

By July 1924, a system of beacons was installed from coast to coast, and airmail service was scheduled for day-and-night operation. The airway was divided into three zones: New York to Chicago, Chicago to Rock Springs, Wyoming, and Rock Springs to San Francisco. A total of 289 flashing beacons marked the route, along with several emergency landing fields.

Flying the Mail

Flying the mail in the early days was not a job for the fainthearted. To survive, a pilot had to be a superb bad-weather navigator and an expert at repairing his own plane.

Flight Conditions One challenge was the Liberty engine, which needed constant maintenance. Pilot Dean Smith sent this terse message after a crash: "On trip 4 westbound. Flying low. Engine quit. Only place to land on cow. Killed cow. Wrecked plane. Scared me. Smith."

The open cockpits presented another challenge. Pilots were constantly exposed to the weather. Even with heavy flight suits, they were often so numbed by the cold that they had difficulty concentrating or working controls. Bad weather and poor visibility made a hard job even harder.

Flight Techniques A pilot blinded by weather had two choices. He could climb in search of clear sky, hoping that he would not become disoriented and tumble into an uncontrollable spin.

Or he could deliberately put the plane into a spin—considered to be a safe way of losing altitude without gaining speed—and hope that he broke out of the mist or cloud cover high enough to recover before the plane hit the ground.

Frank Yeager's ingenious technique also worked. Forced down by fog on the open prairie, he decided to taxi to his destination. Each time he came to a fence, he backed up, hopped the plane over, and continued on his way.

Even in good weather, pilots were always looking for a good spot to land in case they ran out of fuel or had engine trouble. Charles Lindbergh wrote:

> I stare at the earth. There's not a single light on its surface, not a ranch-house window to break this desperate solitude of night. Thousands of feet below I see a huge desert slope . . . Is it veined with creek beds? Is it studded with petrified logs? Are there ridges and cliffs, eroded by rain? The dim light gives little hint of texture . . . Of course I'm lucky to see anything at all—suppose there were no moon, or that the sky were overcast?

In the late 1920s, Swallow Aircraft built mail planes for a number of airlines. United Airlines flew the Swallow pictured below.

Paving the Way for Commercial Aviation

Airmail pilots saw themselves as professionals, not as stunt fliers. This attitude of professionalism led to the formation, in 1928, of the National Air Pilots Association.

Advances brought about by airmail service—such as the airway lighting system—paved the way for long-range, 24-hour commercial aviation. As late as 1930, Europeans dismissed descriptions of the system as typical American boasting—such a long route lit by 36-inch rotating beacons, with well-marked airfields, seemed impossible.

The airmail pilots' on-the-job experience was the strongest argument that better planes, better engines, and better navigational aids were sorely needed. Though the parachutes the pilots received in 1919 were a welcome addition, they were no substitute for instruments that would permit them to fly "blind" in bad weather.

Passenger Airlines

Early Passenger Services in Europe

While Americans focused on mail service, Europeans pioneered the transport of passengers by plane. On August 25, 1919, a British company called Aircraft Transport and Travel inaugurated a civil air service between London and Paris.

Passenger airlines sprang up after World War I. Though many were short-lived, a number of present-day airlines, both European and American, had their beginnings in the 1920s and 1930s.

The flight across the English Channel was uncomfortable but mercifully short. Aircraft Transport and Travel converted the gunner's position in the D.H.4 into seats for two passengers, crammed in face-to-face. The engine, located just ahead of the cabin, pounded so loudly that it was impossible for travelers to converse—even in screaming voices. Instead, they wrote notes to each other and to the pilot, who rode outside in the open cockpit.

New companies in Britain, France, Germany, and the Netherlands (including privately owned KLM in 1919) soon flooded the European market. Too many companies competing for too few passengers caused financial difficulties for the new airlines, and their appeals for government assistance led to the formation of the first national airlines—Imperial Airways (1924), Deutsche Luft Hansa (1926), Air France (1933).

It took Europeans a while to become "air-minded" enough to fill passenger seats, but the record-setting flights of the 1920s helped spur consumer interest, as reflected in the growing number of routes flown by European airlines. In 1919 fixed aerial routes added up to a mere 3,200 miles; by 1925 that figure had climbed to 34,000; and by 1930, to 155,000. Throughout the 1920s, however, state sponsorship remained essential to the operation of the European airlines. In 1929, for example, French airlines received only 10 percent of their income from ticket sales.

Throughout its existence, Aeromarine maintained a perfect safety record. Between 1920 and 1923, the Aeromarine 75 (above), and the company's other models, carried about 35,000 passengers.

Passenger Airlines in the United States

In the United States, the first international passenger service was introduced in 1920 by Aeromarine West Indies Airways in Florida. Flying Curtiss Type 75 flying boats, Aeromarine offered daily service between Key West, Florida, and Havana, Cuba, during the winter season. Instead of facing an eight-hour layover and then taking an overnight boat ride, passengers could set foot in Havana in 75 minutes.

At first, business boomed. In 1921 and 1922, Aeromarine made 2,125 flights and flew 739,047 passenger miles without accident. Encouraged, it opened twice-daily service between Detroit and Cleveland. But in 1924 the company suspended operations—it had not won a lucrative airmail contract, and the profit margin from passenger service was too small.

Between 1919 and 1930, Deutsche Luft Hansa (later Lufthansa) flew five million passenger miles, 22 times the distance to the moon.

The wicker armchairs in the 14-passenger cabin of the Aeromarine 75 helped provide a more comfortable ride.

Throughout the 1920s, attractive government contracts made airmail a higher priority than passenger service. Passengers had to sign a release allowing the airline to drop them off along the route if the company found mail bags to take their place.

Legislation

The Air Mail Act (Kelly Bill) Often called the Magna Carta of civil aviation in the United States, the Kelly Bill of 1925 (amended in 1926) allowed airmail contract holders to be paid up to $3 per pound for the first 1,000 miles of a route, and 30 cents per pound for each additional 100 miles.

The Post Office lost money under this arrangement, but the airlines made profits. In 1929, for example, airmail operators were paid $11,618,000, whereas postal revenues totaled only $5,273,000. The Post Office claimed that the state of affairs was temporary, that the government subsidy of airmail would decline as contractors added passenger services.

The Kelly Bill endowed the postmaster general with enormous power because he determined which companies would receive contracts. Twelve contracts were initially awarded. Among them were Contract Air Mail routes 6 and 7, awarded to Henry Ford and his new aircraft business.

Juan Trippe built Pan Am through sharp maneuvering and his network of influential contacts. By 1929, his company was transporting both passengers and airmail throughout the Americas.

Pan Am: The "Chosen Instrument"

Well-connected and rich, aviation entrepreneur Juan Trippe monopolized U.S. overseas air routes. In 1927, he outmaneuvered two airlines for a mail contract between Florida and Cuba, and formed Pan American Airways. Trippe then extended his reach through Mexico to Argentina, buying out airlines at rock-bottom prices and locking up the Latin American routes. Trippe gained backers in the navy for a Pacific mail route, because it offered an excuse to establish military bases on Midway and Wake Island. The first mail flight to Manila was flown in Martin M-130 Clippers on November 22, 1935, by Pan Am. President Roosevelt cabled the crew: "I thrill to the wonder of it all." In 1939, Pan Am conquered the Atlantic with regular mail service and passenger flights. At that time Trippe's airline received six times more money per mile for carrying mail than any other U.S. airline.

The Air Commerce Act The aviation industry took a further step toward maturity with the Air Commerce Act of 1926, which led to the development of standards for selecting pilots, studies to improve passenger comfort, and increased funds for improving air navigation equipment. The bill required that private and municipal investments fund construction of airports. (In 1933, as part of President Franklin D. Roosevelt's national recovery plan, the government began to pay for airport construction.)

The Air Commerce Act signaled a significant improvement in the passengers' experience in the air. As one observer noted, the government finally accepted that

> flight in aircraft could and should be accomplished without exposing both passengers and pilots to the deafening roar of high powered motors, paralyzing cold, the stupefying effects of the rarefied air at high altitudes, the deadly fumes of carbon monoxide, and the blistering effects of winds of hurricane velocity.

The Major Passenger Airlines

These new regulations and standards made aviation more appealing to U.S. investors. The sudden infusion of government funds, coupled with the lure of mail contracts, brought many companies into the market between 1925 and 1929. Some were manufacturers, like Travel Air Manufacturing Company, formed by Walter Beech and Clyde Cessna in 1925. Its Travel Airs were used by numerous early airlines, but by 1927 the partners had split to create designs under their own names.

United Aircraft and Transport Corporation Gradually, aviation empires formed and put smaller competitors out of business. The first power bloc to appear was United Aircraft and Transport Corporation, on February 1, 1929, and with origins back in 1927 when William E. Boeing, of the Boeing Airplane Company in Seattle, received the airmail contract for the Chicago–San Francisco route.

TWA Another aviation giant took shape as Transcontinental Air Transport (TAT), though its practice of flying passengers during the day and putting them on trains at night lost the company millions of dollars. In 1931, however, TAT merged with Western Air Express to form Transcontinental and Western Air (TWA).

American Another merger was the Aviation Corporation (AVCO), which began with the Chicago-Cincinnati airmail contract and eventually expanded into American Airlines, which came into being between 1928 and 1932. The man responsible for forging the system—the "czar of the airways"—was Walter Folger Brown, postmaster general under President Herbert Hoover. He pushed for passage of the McNary-Watres Act, an amendment of the Kelly Bill that extended his power. Brown used the McNary-Watres Act to force small companies out of business and to build a system of large companies flying multi-engine aircraft. These airlines, he believed, would eventually build both passenger and mail revenues.

The world's first stewardess, Ellen Church (top) went to work for United Airlines in the 1930s. The profession evolved into a glamorous one, though the airlines remained skeptical about women pilots.

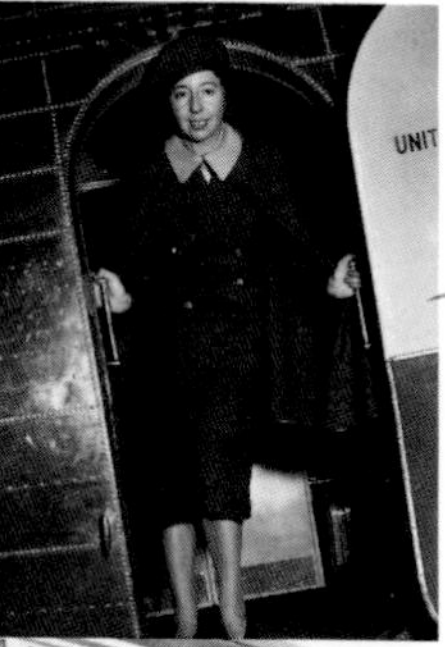

In the late 1920s, trimotor airplanes, which could transport 8 to 16 passengers, were purchased by several airlines and produced by numerous manufacturers, including Fokker, Boeing, and Ford. The Stinson trimotor (above) was operated by American Airlines.

Consolidating the Smaller Lines In the so-called "spoils conferences," Brown established three transcontinental routes, each under a single airline's management. He engineered the merger that resulted in TWA, and he forced the buyouts necessary for American Airlines to stretch across the country. Then, he awarded the middle route to TWA and the southern route to American. United retained the northern route. Other important regional airmail contracts and routes were awarded to Eastern and Northwest.

The small independent airlines fought back. Their complaints led to investigations of Brown's methods, first by journalists and later by Congress. Accused of favoritism, Brown defended his actions but left office in 1933 when Franklin D. Roosevelt became president. Roosevelt canceled all airmail contracts on February 9, 1934. Responsibility for flying the mail fell once again to the army.

This abrupt decision allowed the army only 10 days to prepare for flying airplanes with few instruments during some of the worst weather of the year. By the end of the first week, 5 pilots had died and 6 had been critically injured. A total of 12 died before the mail routes were handed back to private contractors in March.

This time some of the bigger independent airlines—such as Braniff and Delta—won airmail contracts, but the new rules required that all contracts be rebid every year. The major airlines had bid low for their routes, and every one of them operated at a loss. Only with the passage of the Civil Aeronautics Act of 1938 were sound and fair regulations finally established for the industry.

Passenger Planes

In the 1920s, airlines used several different airplane models for passenger service. Among the most popular were two trimotors. Anthony Fokker, transplanted from Holland to New Jersey after the war, produced the Fokker trimotor in 1925. Two significant advances in aircraft design, cantilever wings with no bracing and a metal-framed fuselage, were introduced in this model.

The Ford Tri-motor The next year, Henry Ford directed his engineers to combine the Fokker design with German-developed metal construction to build the Ford Tri-motor, which became known as the famous "Tin Goose." The Ford Tri-motor carried 14 to 16 passengers, had sturdy metal wings, and was powered by an innovative air-cooled radial engine, the Wright Whirlwind. Both the Whirlwind and its competitor, the Pratt & Whitney Wasp engine, were lightweight, reliable sources of power.

Trimotors dominated air transport until more efficient, all-metal airplanes such as the Boeing 247 and the DC-2 became available in 1933 and 1934, respectively. The trimotors were costly to operate, noisy and drafty, and—as demonstrated in the famous crash on March 31, 1931, that killed football coach Knute Rockne—could be fatally weak under stress.

By 1930 new designs were emerging. The Northrop Alpha and the Boeing Monomail represented a growing sophistication in construction and aerodynamic refinements. The Alpha, for example, pioneered the use of an aluminum skin that served as a major structural support, rather than simply covering the plane's frame. To reduce drag, cabin noise, and vibration, the engines of the DC-2 and the Boeing 247 were placed on the leading edge of the wing.

The DC-3 This Douglas design revolutionized passenger transport. The DC-3 was the first profitable passenger plane. Because the DC-3 made money without the encumbrance of mail routes, it bolstered confidence in air travel. Produced between 1936 and 1947, it is still flown today. Together with its military version (the C-47), the DC-3 grossed more than one billion dollars for Douglas.

The Boeing 314, introduced in 1938, became famous for its reliability and luxury on Pan Am's transpacific route between San Francisco and Hong Kong. With room for up to 74 passengers, the interior featured a dining room, a bar, dressing rooms, and sleeping berths.

Douglas DC-3

The Douglas DC-3 made passenger airlines profitable. Able to carry 21 passengers, it began service in 1936. By 1939, DC-3s were transporting 90 percent of the world's airline passengers.

Originally built with 14 berths for overnight travel, the 21-seat DC-3 became the standard model. Adapted for military use as the C-47 and the C-53, it was used in the Berlin airlift. The DC-3, an immortal airplane, is still flown all over the world.

Cowl flaps control the airflow through the engine, to regulate temperature. Engines and even wings could be replaced with a minimum of time and effort.

The cockpit is equipped with an autopilot and two sets of instruments—for pilot and copilot—providing a backup if one set fails.

Two Pratt & Whitney engines deliver 1,200 horsepower, to reach a maximum speed of 230 miles per hour.

Strong landing gear allows the DC-3 to land on short, rough landing strips. Gear retracts to improve streamlining.

Civilian versions carried seats for passengers, or even sleeping berths. Military versions carried soldiers.

The cantilever wings span 95 feet. (To test their strength, Douglas employees drove steam-rollers back and forth over them.)

Flaps allow the DC-3 to land on short runways.

All-metal construction and wind-tunnel-tested aerodynamics produced a streamlined and efficient airplane.

Tail compartment carried luggage or cargo.

The Growth of Commercial Aviation

As the 1920s passed, barnstormers settled down and began to use single airfields as a base of operations. Some aviators conducted scenic tours, carried hunters to remote forest areas, offered flight instruction, ran stock feed to ranchers, and carried supplies to snowbound gold miners. Others herded sheep, bombed ice jams, surveyed for oil, pinpointed archaeological sites, detected forest fires, scouted for tuna, and wrote ads in smoke. In the summer of 1921 a plane was first used to spray insecticide over fields. One company alone dusted 75,000 acres in 1928.

Business executives began buying airplanes, such as trimotors, fitted with desks and dictation machines, at a fast clip. By 1930, one third of the civil aircraft purchased in the United States transported business people around the country. Cargo service also grew rapidly, topping four million pounds by 1930.

Classic Lightplanes

After World War II, the aircraft industry geared up to sell planes to pilots who had learned to fly during the war. Companies like Aeronca, Beech, Cessna, and Piper succeeded in bringing recreational flight to many middle-class Americans. Small planes like the Cessna 150, the Aeronca Champion, the Beechcraft Bonanza, and the Piper Cub enabled farmers to check crops, business people to attend conferences, families to vacation afar, and enthusiasts to joyride.

Introduced in 1936, the popular Piper Cub is a two-seat, high-wing lightplane, the first in a Piper line that includes the Super Cub and the Cub Special.

The Crash of the *Hindenburg*

Few disasters have evoked such immediate horror and lingering fascination as the explosion of the great German airship *Hindenburg.* This 804-foot dirigible, emblazoned with the Nazi swastika, had crossed the Atlantic several times, without even a minor accident, since its first flight in 1936. Passengers paid exorbitant fares ($400 from Germany to New Jersey) to travel in quiet luxury between 1,000 and 3,000 feet above the ocean, at a speed of about 75 miles per hour. The *Hindenburg* was equipped with a dining room, a lounge with an aluminum grand piano, a bar, promenades, private cabins for passengers, and a heating system.

On May 6, 1937, the airship had enjoyed an uneventful passage from Frankfurt, Germany. Then, while preparing to moor at Lakehurst, New Jersey, the *Hindenburg* was ripped by an explosion that sent flames racing through the airship. Eyewitnesses heard screams over the roar of the inferno. Of the 97 people on board, 35 died as the burning ship crashed to the ground.

At the time, the disaster was attributed to "static electricity discharge." Perhaps flammable hydrogen had leaked from the gasbag and ignited. Rumors that a bomb had been planted by anti-Nazi activists long persisted but were never proved. Whatever the cause, the era of passenger airships was over.

World War II

A far cry from the wooden biplanes of World War I, the fighters and bombers of World War II had greater speed, came in greater numbers, and had a far deadlier effect. Again, war sped the development of aircraft, including the first jets. Ushered in by Germany's dive bombers in Europe, the war was brought to a grim finale with the explosion of two atomic bombs in the Pacific.

A pilot's World War II memorabilia include medals, flight gear, and instruments.

When Willy Messerschmitt first presented the Me 109 to a Luftwaffe official, he was told, "This will never be a fighting airplane. The pilot needs an open cockpit. He has to feel the air to know the speed of an airplane." The fighter's performance soon proved him wrong.

Air Power at the Outbreak of the War

In September 1939 the screaming of Stuka dive bombers marked the beginning of World War II. Czechoslovakia had already fallen; then Poland was attacked—the German war machine cut through Central Europe swiftly and decisively.

Blitzkrieg

Horrified, the rest of Europe watched the Germans unveil a new kind of mechanized war called *blitzkrieg* (meaning "lightning war"), in which air power played a key role. Bomber attacks on airfields and industrial centers preceded ground operations. Then, as huge armored divisions rolled into enemy territory, fighters ranged ahead of the ground troops to disorganize and demoralize the enemy. Germany's air force was called the Luftwaffe ("air weapon").

The Polish air force did not stand a chance against blitzkrieg: by October, 90 percent of its planes and 70 percent of its air crews had been annihilated. By April 1940, air forces in Denmark and Norway had been overcome. Five days of lightning war in May forced the surrender of Holland; Belgium followed only thirteen days later.

Germany's Planes

The Stuka Early in the war, the Stukas attacked roads and railways and terrorized civilians with the eerie whistle they emitted in dives. The powerful Junkers Ju 87 Stuka, which could achieve a speed of 235 miles per hour, led Germany's invasion of Poland.

The Messerschmitt 109 Germany's premier fighter in 1939, the Me 109, carried machine guns and cannon, a brutally effective combination. Germany also fielded the Me 110 fighter, which performed poorly at the beginning of the war but later excelled as a night fighter.

The Battle of France

France felt safe behind the Maginot Line, confident that the German army could not cross that impressive fortification along France's eastern border—three lines of reinforced concrete, blockhouses, and forts that bristled along the German frontier from Belgium to Switzerland. The Allies positioned their mobile forces in the north, along the Belgian border.

The French were vulnerable to air attack. After the nationalization of defense industries in 1936, the French aircraft industry had forfeited its former excellence. Well-established companies had disappeared, and internationally known designers such as Henri Farman were no longer active.

As the Germans arrived on the border, the French air force faced the enemy's 1,300 long-range bombers, 400 dive bombers, 1,200 fighters, and 650 reconnaissance aircraft with only a few hundred airplanes of its own. Even the best of them, the Dewoitine 520 and Morane-Saulnier MS 406, were slower than the Me 109s and too few in number. When fighting began, they suffered heavy losses.

The Curtiss Hawk 75 The French made frantic requests for aircraft to the United States and to Britain. An American fighter, the Curtiss Hawk 75, reached France in significant numbers. Though outnumbered and outperformed, Hawk 75s shot down 311 Luftwaffe planes.

The Hawker Hurricane The most notable plane the British sent across the English Channel was the Hawker Hurricane. A low-winged monoplane with an enclosed cockpit, eight-gun armament, and a Merlin engine, it represented a huge leap forward for the Royal Air Force (RAF). Fabric and wood biplanes had dominated the force a mere five years earlier. Even with the Hurricane, however, France's fleet was vastly outnumbered by the Luftwaffe.

Ground crews prepare for an air raid by mounting bombs under an airplane's wings (below). The invasion of Poland, leading to the swift decimation of the Polish air force by German bombers, including Ju 87 Stuka dive bombers, led Britain and France to declare war against Germany in September 1939.

The Messerschmitt 109 and the P-51 Mustang

The Rolls-Royce Merlin engine reliably powers a four-blade propeller, for a maximum speed of 440 miles per hour. Drop tanks holding 1,150 gallons were added to extend its range to escort long-range bombers.

Designed by Willy Messerschmitt, the Me 109 was employed by the Germans throughout World War II. First put to the test in the Spanish civil war, the airplane was remarkable for swift climbs, fierce dives, and maneuverability.

The Me 109 was used extensively in the Battle of Britain, where it was challenged but not overwhelmed by the Supermarine Spitfire.

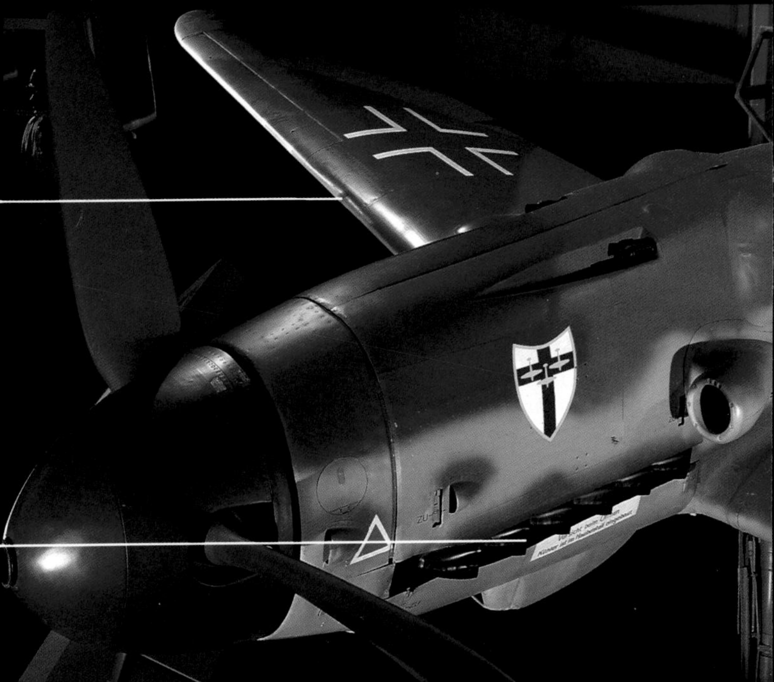

The standard armament of two machine guns and one 20-millimeter cannon could be augmented by two below-wing guns. The Me 109 sometimes carried bombs as well.

The Daimler-Benz liquid-cooled engine enables the fighter to reach a maximum speed of 385 miles per hour.

The pilot is shielded by a "teardrop" sliding canopy.

More than 15,000 P-51s were built starting in 1940.

Armament consists of six machine guns.

The streamlined wings and fuselage reduce drag, making the Mustang an unusually fuel-efficient fighter.

The North American P-51 Mustang was the leading long-range fighter of World War II. Swiftly designed and prototyped in 1940, it performed impressively in both the European and Pacific theaters of the war.

In 1942 the Mustang was fitted with Rolls-Royce Merlin engines—all built by Packard in the United States—and its speed increased to a brisk 440 miles per hour. Mustangs served in various air forces until the 1970s.

Between 1936 and 1945 an estimated 35,000 Me 109s were built. Some versions remained in use in several countries until the mid-1960s.

The all-metal construction features low, cantilever wings and retractable landing gear.

In mid-1938 Britain owned only three Supermarine Spitfires (above). But by the outbreak of war in September 1939, more than 600 Hurricanes and Spitfires had been produced to face the threat of German invasion.

The Royal Air Force Arrives

The RAF landed in France in September 1939 only to find soggy grass airfields and no hangars at all; the air crews slept in barns and shaved in streams. The Battle of France was the first taste of war for many RAF pilots. And they flew in close formations exactly as they had been taught in flight school.

In contrast, the Luftwaffe had primed itself fighting in the Spanish civil war, where the pilots learned to fly in loose pairs rather than tight formations. They poked fun at the Allied airplanes, which they contemptuously described as "bunches of bananas ready for the picking." They later changed their tune.

The RAF's loss rate was high. One squadron lost 17 out of 20 pilots; a bomber squadron of Bristol Blenheims lost 11 of 12 aircraft in one attack. According to British Air Chief Marshal Sir Hugh Dowding, "I saw my reserves slipping away like sand in an hourglass."

The Germans Push Through

In May 1940 the Germans bypassed the Maginot Line and cut through the Ardennes forest in southern Belgium all the way to the English Channel. The Allied forces had moved into northern Belgium and the Netherlands—which quickly fell to the Germans, leaving the Allies cut off from France. Abandoning equipment and matériel, they retreated to the beaches of Dunkirk, a small port on the northern edge of France. Here the British Royal Navy laid plans to rescue the Allied soldiers.

The Evacuation from Dunkirk

As the Allied ground forces backed up to the Dunkirk beaches, the Luftwaffe prepared to deal a death stroke. Reichsmarshall Hermann Göring, leader of the Luftwaffe, bragged that his planes could sink the rescue ships and destroy the troops streaming to Dunkirk.

Between May 26 and June 4, 1940, the RAF Fighter Command flew 300 sorties a day over the embarkation area at Dunkirk, losing 106 aircraft and 80 pilots. But they destroyed 132 Luftwaffe planes. Far from destroying the Allies, the Luftwaffe allowed 340,000 Allied troops to escape from a triangle less than 1,000 miles square. Göring had miscalculated—the Allied ground forces were brilliantly defended from the air.

The Supermarine Spitfire The legendary Supermarine Spitfire first faced off against the Messerschmitt 109s and 110s above Dunkirk. More advanced than the Hurricane, the Spitfire gained the Germans' grudging respect. It outmaneuvered the Me 109 in skirmishes at high altitudes and was modified throughout the war to keep up with improvements in German aircraft.

In World War II lore, the Spitfire has often been glorified and the Hurricane ignored, their differences likened to those between a greyhound and a bulldog, a racing car and a truck. German pilots bested by the British also developed "Spitfire snobbery" and invariably claimed they had been shot down by a Spitfire. But squadrons that flew Hurricanes called it a "go anywhere, do anything plane, a defensive fighter, a workhorse."

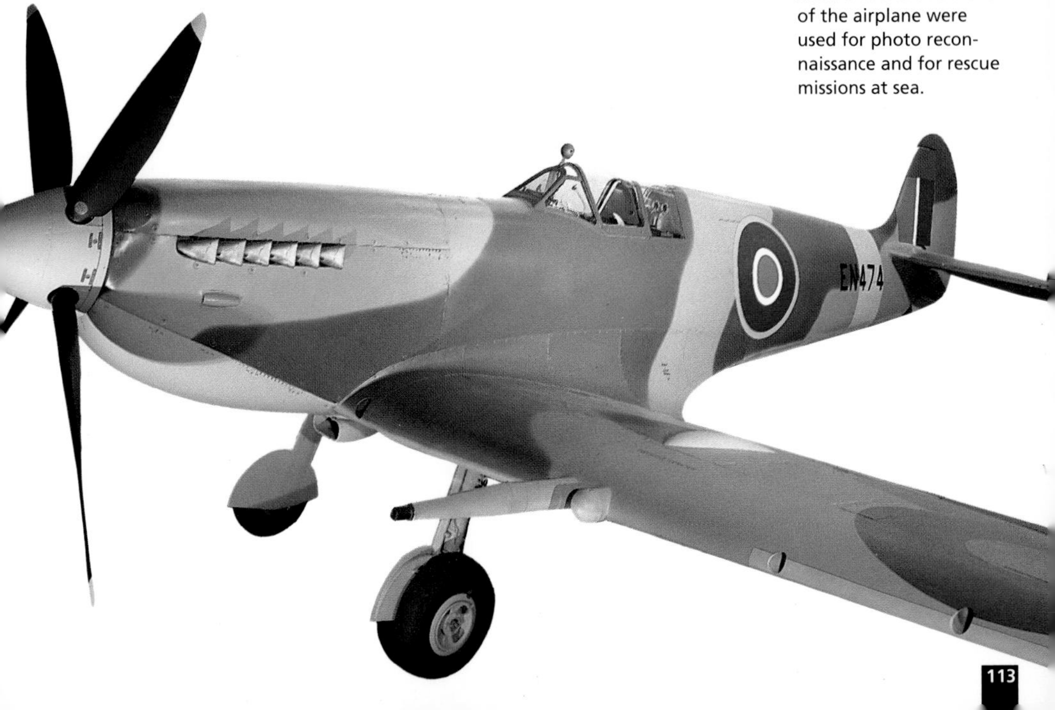

The Supermarine Spitfire was produced and refined throughout the war. Armed with eight machine guns, it was a fierce opponent in the skies. Different versions of the airplane were used for photo reconnaissance and for rescue missions at sea.

The German Junkers Ju 88 was a twin-engined dive bomber. Although its normal load capacity of 3,968 pounds was considered rather light, it was a successful night fighter and bomber. More Ju 88s were built during the war than any other German bomber.

Setting Up Operation Sealion

At Dunkirk, the Luftwaffe had failed for the first time—and against a retreating enemy fighting on foreign territory with few resources. Instead of taking this as a warning, Göring assured Hitler that he could destroy the RAF on its home turf in a matter of weeks. They began planning an air assault on Britain that would culminate in an invasion called Operation Sealion.

The Luftwaffe's Strengths—and Weaknesses Though the Luftwaffe was formidable (by the end of July 1940, it numbered 1,200 bombers, 280 Stukas, and 980 fighters, based on fields from Norway to Brittany), the Germans had certain vulnerabilities. They lacked a radar system on the English Channel coast, and their star fighter, the Me 109, had a limited range and could not escort bombers over Britain without running low on gas.

There were also problems with the German bombers. In ideal conditions, the Stuka dive bomber was devastating, but it was highly vulnerable to antiaircraft fire and to fighters, and was eventually withdrawn from the battle. The Junkers Ju 88A bomber was fast and maneuverable for its size but lacked sufficient defensive armament. The Luftwaffe simply did not have the type of bomber needed to destroy the RAF bases in Britain.

British pilots run to their airplanes during the Battle of Britain—control of the air turned out to be decisive in preventing the invasion of Britain.

The Battle of Britain

According to Hitler and Göring's plan, the Luftwaffe would set the scene for Operation Sealion by attacking across the channel to demoralize and weaken the British forces. Then, 13 divisions of ground forces would land along a 225-mile stretch of British coast. But the ground troops would wait until the RAF was crushed so that it could pose no threat to the crossing. Hitler gave Göring until the early fall of 1940 to accomplish this task.

Early Skirmishes

The German attacks commenced in July. Initially, the Battle of Britain looked like a walkover for Germany. In one early engagement, nine Boulton-Paul Defiant fighters were attacked by 109s over the English Channel. The British fighter squadron was nearly wiped out in its first engagement, with only two planes surviving.

The Luftwaffe definitely had the advantage in numbers. In July, the RAF in southern England could put up only 600 planes against 1,500 German aircraft. Seemingly invulnerable, the Me 109s zoomed unopposed across England's skies at 25,000 feet.

But the Luftwaffe was stumped when RAF pilots ignored the fighters. The Me 109s were challenged only when they accompanied bombers. Then, when the RAF fighters did engage, the Luftwaffe pilots received another shock. Luftwaffe losses edged toward 300 planes, and the Spitfires and Hurricanes kept coming.

Test Pilot Hanna Reitsch

In the early 1930s, Hanna Reitsch gained an international reputation as a glider pilot. The Luftwaffe called on her to test military aircraft during the secret rebuilding of Germany's air force in the 1930s.

After testing a huge Messerschmitt glider capable of carrying a tank and 200 soldiers, Reitsch agreed to fly the Messerschmitt 163 experimental rocket plane. The Me 163 could achieve a speed of more than 500 miles per hour and a height of over 30,000 feet in just over two minutes.

Reitsch's first four tests were successful, but on the fifth, the plane's undercarriage jammed. Reitsch struggled to land the plane rather than bail out of a valuable machine. For her efforts she received a serious head injury—and the Iron Cross, from Hitler.

When she recovered, she was horrified to find Göring ready to put the defective Me 163 into production: "It became only too clear that Göring did not wish his comforting illusions to be disturbed."

Reitsch also flew Nazi leaders on missions to combat zones and occupied territories. Her offer to form a squadron of women fighter pilots was turned down.

Britain's Chain of Command

Besides flying over home ground, the RAF had other advantages, including an outstanding chain of command headed by Sir Hugh Dowding, whose organizational talents extended from developing an aerial strategy to insisting that the Hurricanes and Spitfires be equipped with bullet-proof windshields.

The minister of aircraft production, Lord Beaverbrook (a Canadian-born newspaper tycoon) sped up manufacturing—even beseeching British housewives to send him pots and pans to build Spitfires. Whether he actually used any of this kitchenware, he certainly produced a lot of planes. In February 1940, 141 fighters were manufactured; by June, 446 fighters a month were rolling off the lines.

Responses to German air attacks were directed from this observation room—staffed by RAF personnel—the nerve center of British air defense.

The Me 109, first demonstrated at the 1936 Berlin Olympic Games, was the premier German fighter early in the war.

"Superbinoculars"

A nasty surprise lay in store for the Luftwaffe pilots arriving over England—fighters, ready and waiting. The new British radar system—which the German pilots called "superbinoculars"—provided ample warning of approaching Luftwaffe aircraft to the British Fighter Command.

Called the Chain Home system, these radar posts were equipped with 350-foot steel towers to guard the British coast from southern England to the tip of Scotland and could identify aircraft at a height of 15,000 feet from a distance of 40 miles.

The Chain Home Low stations focused on low flights and identified fighters coming across the English Channel at sea level. Luckily for Britain, the Luftwaffe never recognized the importance of these stations. To them, it seemed unlikely that Fighter Command would place such important resources above ground and near airfields, without even sandbags to protect them.

The radar towers were obvious bombing targets, but the British repaired damage quickly or brought in mobile units. Completely stymied, Göring proclaimed, "It is doubtful whether there is any point in continuing the attack on radar sites, in view of the fact that not one of them attacked so far has been put out of operation."

The Observer Corps Supporting the radar was a large Observer Corps. These men and women tracked the activities of hostile and friendly forces over Britain. This information was fed into sector and group command rooms and finally into Fighter Command. The central feature of the Operations room at Fighter Command was a huge map of Britain spread out on a table. Markers representing the attacking and defending squadrons were moved around the map with sticks resembling pool cues.

Fighter Planes in the Battle of Britain

The RAF and Luftwaffe fighter planes were a fairly even match, each with particular advantages. The Me 109 could outdive the Hurricane and Spitfire with ease, though the Spitfire could match it in a climb.

The Merlin engine of the Hurricane and Spitfire tended to cut out momentarily when pushing over into a dive, and in those few lost seconds a German fighter could break off the engagement. The British fighters found their advantage in low-altitude maneuvering and turning.

Fighter Pilots

A normal day for a fighter pilot was pretty much the same on both sides of the English Channel. Up at 3:30 A.M., pilots were on call until 8 P.M., running to their planes as ordered and then dozing between flights. Dogfights were short: a fighter had at most 15 seconds of ammunition to fire before it was left defenseless.

Unlike the airplanes of World War I, these fighters were equipped with radio communications. Pilots warned each other of approaching enemy aircraft and offered congratulations on scoring a victory. They talked in their own lingo. Among British pilots, to "scramble" was to take off, to "pancake" was to land. After returning from flights, they discussed details of the plane's performance with an "erk," a member of the ground crew. Each hoped to avoid a "prang," a crash.

Pilots found the work grueling: "After eight scrambles in a day, you came to write up your logbook . . . and you just couldn't remember beyond putting down the number of times you'd been up . . . you couldn't remember at all . . . I had nightmares about blazing planes crashing all around me."

Aerial combat was dramatically captured through the gun camera of an American fighter pilot. From left to right, a German Me 109 is shown with an attacker close on its tail. After his plane's belly tanks explode, the German pilot desperately attempts to gain altitude. The crippled aircraft actually crashed into a snow bank.

Top American ace Major Richard Bong broke aviation records when he shot down 40 Japanese aircraft in the war in the Pacific.

August 13–24, 1940: Germany's New Tactics

August 13, 1940, was to be *Adlertag*, the "day of eagles"—the day Britain was to fall. The Luftwaffe attacked, flying 1,485 sorties, and lost 46 aircraft. British Fighter Command flew 700 sorties and lost 13 aircraft. The worst-hit airfield was back in operation in ten hours.

Between August 13 and 18 the Luftwaffe stepped up the pressure, attacking 34 airfields and 5 radar stations, some of them repeatedly. Then, desperate for results, Göring changed the Luftwaffe's methods.

German Planes Change Roles First, the Stukas were withdrawn from battle. In one six-day period, 41 Stukas had been shot down; they were clearly too vulnerable. The Me 110s, never designed to be escort fighters, had failed in that role; during the month of August, 120 of them were destroyed. Instead of staying with the bombers, the Me 110 pilots began retreating into protective circles. Göring forbade the use of this tactic but could not withdraw the 110 due to a shortage of the superior Me 109s.

That airplane, his most effective fighter, was pressed into service as a bomber escort, guarding ungainly Dornier and Heinkel bombers through the British skies. As few as 120 fighters were often asked to cover a bomber formation 40 miles long.

A New Target: London The RAF Fighter Command was actually weakening—but the Germans didn't realize it. More British planes were being destroyed than could be produced; more pilots were being killed than could be trained. Then—unexpectedly—a grace period allowed the RAF to regroup.

On August 24, a few Luftwaffe bomber crews mistakenly dropped bombs on central London, thinking they were attacking oil tanks on the city's outskirts. Prime Minister Winston Churchill immediately responded by attacking Berlin, after which Adolph Hitler ordered the destruction of London. This exchange would radically alter the course of the Battle of Britain.

The French military decoration, the croix de guerre, was awarded to members of the Allied services in both world wars in appreciation of gallantry.

Beginning on September 7, the Luftwaffe stopped attacking fighter airfields and focused on London, an attack known as the Blitz. For 67 out of 68 nights an average of 163 bombers per night harassed the city. The Luftwaffe dropped nearly 14,000 tons of high explosives and over 12,000 incendiary canisters on London. These attacks damaged factories, communications, and utilities, but war production was not seriously affected.

The End of Operation Sealion

By September, with German aircraft taking increasingly heavy losses over Britain, Hitler realized that the charge he had given the Luftwaffe—to clear the RAF from the skies—could not be carried out. Reports of German successes had been grossly inflated. "You have apparently shot down more aircraft," he complained to Göring, "than the British ever possessed."

In light of this failure, Hitler postponed Operation Sealion in mid-September. Then, on October 12, he renounced all intentions of landing on British soil in 1940. The battle was over.

Pearl Harbor: The War in the Pacific Begins

In preparation for a raid, this fighter pilot has donned full flight regalia—including a flying helmet with oxygen mask. Flying gear included wool-lined flying boots, long woolen socks worn over silk ones, inflatable life jackets called "Mae Wests," and a parachute harness.

At 5:00 A.M. on December 7, 1941, Japanese pilots gathered in front of Shinto shrines, a standard feature on Japanese aircraft carriers. They ate a special breakfast of tiny red beans with rice, then returned to the briefing room for one last look at the charts and maps. Flight leaders warned them not to touch their radios until the attack began. Then the Mitsubishi Zero fighters, "Val" dive bombers, and "Kate" torpedo bombers roared off the flight deck. As they reached Oahu, the clouds cleared.

Close to the U.S. naval base at Pearl Harbor, flight leader Mitsuo Fuchida fired a flare—the prearranged signal for Japan's torpedo planes to attack first, followed by dive bombers, indicating that the base had been taken by surprise. Fuchida thought the fighters had not seen the flare and fired another one—but two flares signaled that the U.S. base had not been taken by surprise, a scenario in which the dive bombers were to attack first. Confused, all the pilots attacked at once.

The surprise was so complete that the bungling of the flares had no effect. Pearl Harbor's watch officers assumed the planes were a flight scheduled to arrive from San Francisco. Men on board ship waved to the pilots, assuming they were visiting Russians. Cornelia Fort, a flight instructor, had to grab the controls away from a student as a fighter zoomed across their path. Angrily trying to identify the craft, she froze as she saw that "painted red balls shone on the wings."

The Aircraft Carrier in the Pacific Theater

Between 1941 and 1945, aircraft carriers were a decisive factor in battles in the Pacific. Called "flattops" by U.S. forces, they were capable of carrying up to 100 aircraft and 3,500 personnel. A typical aircraft carrier was 820 feet long at the waterline.

These floating airfields stayed at sea for weeks at a time, supported by fuel tankers and supply ships. Unlike battleships, aircraft carriers were not protected by armor plates, and they carried explosive plane fuel. Susceptible to attacks from dive bombers, torpedo bombers, and submarines, the carriers were defended with antiaircraft guns. After an attack, repair of the flight deck was the top priority, so that aircraft could be launched or landed as soon as possible.

A crash barrier, a flexible net raised above the deck, saved many lives, as damaged aircraft returning from battle skidded to a halt in the barrier.

In the last year of the war in the Pacific, the American fleet came under attack from kamikazes, Japanese pilots who volunteered to crash their explosive-laden planes into Allied warships. Even after Japan surrendered, on August 15, 1945, Admiral William Halsey suspected that kamikazes might disobey orders and continue to attack, so he ordered "any ex-enemy aircraft attacking the fleet to be shot down in friendly fashion."

At 7:53 A.M., before a single bomb had fallen, the flight commander radioed the Japanese fleet that the attack was successful. When the smoke cleared, 2,400 Americans were dead and 1,500 wounded. The toll on the fleet was high: 4 warships sunk, 14 ships heavily damaged, and 233 airplanes destroyed.

Attack on the Philippines

The United States had expected Japan to attack its base in the Philippines, not Pearl Harbor. General Douglas MacArthur had every reason to believe the next attack would be on his forces at Clark Field near Manila, but only a few U.S. fighters were able to get into the air when the Japanese Zeros attacked on December 8. They had no warning because the radar set at Clark Field was broken. Four Curtiss P-40E Warhawk fighters were destroyed during takeoff, and 14 others were eliminated before they reached the runway. Most of the B-17 bombers were destroyed.

For the next three months, the Americans fought a gallant but doomed battle of attrition against the Japanese. No mail, no supplies, and no reinforcements ever reached them. Each Japanese attack numbered 30 or more Zeros against a defending force of a few P-40s.

The Japanese invaded and occupied Luzon, the largest island of the Philippines, on December 22, 1941. The remaining force of 15,000 Americans and 45,000 Filipinos halved their food rations. Air cover shrank to a total of nine P-40s. By April, the defenders were living on rice and lizards. Only two airplanes remained when they surrendered.

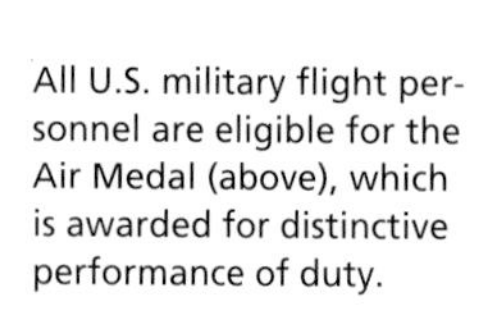

All U.S. military flight personnel are eligible for the Air Medal (above), which is awarded for distinctive performance of duty.

Fighters in the Pacific

The Mitsubishi Zero The Mitsubishi Zero fighter gave the Japanese air superiority over New Guinea and the Solomon Islands early in the conflict. The Zero could climb faster, fly higher, and outmaneuver U.S. fighters.

The Japanese pilots ran rampant over squadrons that used tactics developed against the German Me 109s. The number of green U.S. pilots was another weakness. Even the Spitfires flown by the Australians (who fought beside the U.S. forces) had difficulty with the Zeros. Clearly, Allied forces had to learn how to fight Japan's planes.

The Curtiss P-40 Warhawk Though less maneuverable than the Zero, the Warhawk was a reliable plane that could achieve a speed of 350 miles per hour—a match for the Zero in level flight. Americans eventually learned to attack from a dive, then shoot and break away.

Doolittle's Tokyo Raid

In April 1942, Lieutenant Colonel James "Jimmy" Doolittle spearheaded an act of defiance that boosted Allied morale. At the time, the Japanese were raising their flag over more and more captured territory. The U.S. armed forces had been caught embarrassingly unprepared at both Pearl Harbor and in the Philippines.

Launched from a carrier 750 miles off the Japanese coast, Doolittle led a flight of 16 B-25 Mitchell bombers against Tokyo. The damage the planes caused was slight—they were too heavily loaded with fuel to carry many bombs.

But at a time when all war news was bad, the Doolittle raid received tremendous coverage. Though the crews had to ditch their planes in China, most of the men managed to make their way back and were given a hero's welcome.

Assembled on the deck of the U.S. aircraft carrier *Hornet* (below) are the 16 B-25 Mitchell bombers awaiting dispatch on the mission known as "Doolittle's Tokyo Raid." The crews, posing for a preraid photo (inset), were briefed as soon as the *Hornet* was at sea. Doolittle was awarded the Congressional Medal of Honor for his leading role in the raid.

The Grumman F4F Wildcat The other principal American fighter in the Pacific at the time was the Grumman F4F Wildcat. Unlike the Warhawk, the Wildcat was not as fast as the Zero, but Wildcat pilots knew they were flying a tough machine. It could take heavy punishment and still stay in the air, and was equipped with cockpit armor that protected its pilots well.

The Battle of the Coral Sea

The Battle of the Coral Sea was the first battle fought entirely in the air; the carriers on either side never made visual contact or exchanged fire. The Japanese attempted to take Port Moresby, in New Guinea, by sea. Damage was inflicted on both sides, including the loss of the American carrier *Lexington.* Though indecisive, this battle of May 1942 prevented the Japanese forces from gaining ground in New Guinea.

The Battle of Midway

On June 4, 1942, Japan attacked Midway Island with four carriers and supporting ships. But the U.S. Navy had a decisive advantage—it had broken the Japanese code and was ready for the attack. Planes from the American base on Midway conducted the initial defense.

Admiral Chuichi Nagumo did not realize that three U.S. carriers were waiting 350 miles northeast of Midway. The Japanese

A fighter to contend with, Japan's Mitsubishi Zero combined agility, speed, and range. Nicknamed "Zeke" by the Allies, it participated in every major action of the war in the Pacific, including Pearl Harbor.

The Flying Tigers

From December 1941 until July 1942 General Claire Chennault led a small group of 100 volunteer pilots, nicknamed the Flying Tigers, who fought the Japanese in China. In their six months of service before being incorporated into the U.S. Army Air Force, they claimed an amazing 300 victories over the Japanese, losing only 50 planes and 9 pilots.

The Flying Tigers flew heavy P-40 Warhawks that were sluggish in climbing and hard to maneuver. So Chennault devised tactics to play up the P-40's strengths: good speed in straight-away flight and dives. Outnumbered by lighter Japanese planes, the Flying Tigers fought in pairs—diving, shooting, and breaking away, always avoiding one-on-one duels.

decimated the first flight from these carriers—TBD Devastator torpedo bombers that attacked at sea level—without realizing their source. In battling the Devastators, the Japanese used up fuel and ammunition and pulled their air cover down to sea level. Then the Wildcats and the dive bombers, which attacked from high altitude, arrived over the fleet.

By the time Nagumo realized he was fighting three carriers, three of his four carriers had been sunk. The fourth soon followed. In a carrier-based war, such a blow was devastating. Along with his ships, Nagumo lost 250 planes and crews, 3,000 sailors, and many of his most skilled pilots.

In August, U.S. forces began a campaign to wrest Guadalcanal from the Japanese. After long, bitter fighting, they were able to oust the Japanese in February 1943. The Japanese never regained the offensive in the Pacific.

U.S. Airplane Production in World War II

The U.S. aircraft industry went to work long before the government declared war. By the late 1930s huge orders from Britain and France had started the airplane production lines rolling.

For example, in 1940 the British went to North American Aviation looking for a new fighter for the RAF. A prototype for the Mustang was flying within 117 days, eventually providing the ideal escort for U.S. bombers in Europe.

The U.S. aircraft industry produced planes at a mind-boggling rate. In 1939, 2,200 military aircraft were produced. By 1944, annual production had reached 100,000. A single plant, the Ford facility at Willow Run, produced 4,476 B-24 bombers in 1944 and 1945.

U.S. companies concentrated on producing 18 basic fighter and bomber models.

Women played a cru role in the war effort Besides military servi millions of women en tered the workforce the first time to fill v slots in defense facto Many became skilled chanics and welders; these women are con structing self-healing tanks.

These Republic P-47s part of the estimated $45 billion that the U government spent fo craft in the war.

Bombing Strategies in the European Theater

Between 1940 and 1944 a number of bombing strategies were developed by both German and Allied commands.

Precision Bombing

Germany's bombing of London, intended to lower civilian morale, had virtually no effect on war production. But Britain's Bomber Command unwisely pursued the same tactic, devoting substantial resources in the precision bombing of German targets.

The blatant failure of Britain's early night raids on Germany was also ignored. On December 16, 1940, for example, a fleet of British bombers—Wellingtons, Whitleys, Hampdens, and Blenheims—attacked Mannheim. Though the crews reported the center of town to be in flames, reconnaissance photographs later showed that most of the bombs had fallen wide of their targets.

Bomber crews did not yet have reliable navigational aids. Even in daylight, RAF officers estimated that crews flying above cloud cover could not reliably come within 50 miles of a target. Night bombing rarely achieved much accuracy. Even in favorable weather conditions, night crews could be expected to drop only 2 to 4 bombs out of 100 on a target the size of an oil refinery.

Still, throughout the spring and summer of 1941, Bomber Command sent crews against specific targets at night. Throughout June and July, not more than one in five planes arrived within five miles of its target.

Area Bombing

Later that year, when the failure of precision bombing could no longer be ignored, British strategy changed to area bombing. The goal of destroying Germany's industry would be met by sending fleets of up to 1,000 airplanes to unload tons of bombs on a given target. Bomber Command felt certain that six months of this treatment would break the resolve of the Germans.

Once again, the results proved otherwise. Between August and October, Bomber Command flew 8,466 sorties, with negligible effects on production or morale.

When the U.S. Eighth Air Force, based in Britain, became active in August 1942, General Henry "Hap" Arnold and his commanders believed that their heavy B-24 and B-17 bombers could defend themselves in large formations. Early raids against aircraft factories and oil fields proved them wrong. In a mere seven days, the Eighth lost 150 bombers.

Despite its size, the elegant lines of the B-17, or Flying Fortress, made it an attractive subject of recruiting posters for the U.S. Army Air Force. In the United States and Britain, the B-17 is still the object of much affection.

Bombing Around the Clock

In January 1943, Churchill and President Franklin Roosevelt agreed to open a combined bomber offensive against Germany in preparation for a cross-channel invasion. The goal of the offensive was to destroy the Luftwaffe and Germany's sources of oil. The RAF would attack by night, U.S. forces by day.

Never before had air power been the vehicle for such a massive, concerted attempt to destroy the industrial, economic, and military base of a modern nation. Yet, initially, the combined forces had no great success.

The B-17 and B-24 Bombers Called the Flying Fortress, Boeing's B-17 was prominent in Allied bombing missions from late 1942 until the end of the war. It deserved its nickname: the B-17 could carry up to 17,600 pounds of bombs and was armed with twelve machine guns for defense.

Consolidated Aircraft Corporation built the B-24 Liberator bomber as an improvement on the B-17. Flown by U.S., British, and Canadian forces, it is perhaps best known for bombing the Ploesti oil fields in Romania in 1943, an ill-fated raid in which 54 B-24s were lost.

Among the new design features of the Consolidated B-24 Liberator was a retractable, tricycle-type landing gear. Although the Liberator served in Europe, it saw more action against the Japanese in the Pacific.

The Allies dropped 1.5 million tons of bombs on Germany during World War II.

Allied Fighter Escorts To reach the heart of Germany with bombs, the Allies clearly needed a long-distance escort fighter. Yet the principal U.S. fighter in Europe, the Republic P-47 Thunderbolt, affectionately called the "Jug," did not have the range to escort bombers deep into German territory.

It was the North American P-51 Mustang that took on the role with flair. The P-51 was designed, built, and flown within 117 days. Equipped with a Merlin engine, the Mustang could fly at 440 miles per hour, with a ceiling of 41,900 feet and a range (with drop tanks) from Britain to Berlin and beyond. Total production of the Mustang was 14,490 planes, whose prime mission was to escort the "Big Brothers" (the bombers) and establish Allied air superiority over Germany.

On long, high-altitude missions, sometimes fighter pilots experienced problems due to *hypoxia* (oxygen starvation). Oxygen systems could develop leaks, and sometimes the oxygen mask did not fit the pilot well. This caused some pilots to black out—and their planes to fall in near-vertical dives. One pilot wrote of the horror of watching a friend crash into the North Sea: "I followed him down, yelling at the top of my voice, imploring him to wake up."

The P-51 Mustang was faster than any rival German fighter. The P-51 also had two to four times the internal fuel capacity of competitive aircraft.

Cockpit of the Boeing B-17G

Gyro compass

Radio compass

Altimeter

Airspeed indicator

Pilot's seat

The B-17G Flying Fortress, Boeing's first heavy bomber, required a crew of ten. The cockpit was usually occupied by a pilot, a copilot, and a flight engineer.

New Fighter Planes for the Axis

Though well suited to the escort role, the Mustang faced a tall order. It flew against some superb enemy fighters.

The Focke-Wulf 190 The Me 109 had already proved its worth in the Battle of Britain. Since then, the Focke-Wulf 190 had also come into its own. Allied pilots marveled at the German engineering of the plane: light on the controls, an easily maintained engine, and a wide, sturdy undercarriage. Pilots never forgot their first glimpse of a 190: "Never had I seen so beautiful an airplane. A rich, dappled blue, from a dark, threatening thunderstorm to a light sky blue. The cowling is a brilliant, gleaming yellow. Beautiful, and Death on the wing."

The Macchi C.202 Folgore Italy's best fighter, the deft C.202 could speed through the skies at almost 375 miles per hour. Dogfights with the C.202 were no picnic for an Allied pilot: "One moment I am thrust down into the seat in a tight turn; the next I am upside down, hanging in the safety harness with my head practically touching the canopy roof and the guts coming up into my mouth. Every second seems like a lifetime."

Air Supremacy for the Allies: 1944

By 1944 the effectiveness of Allied bombers was increasing. Better fighter escorts and the introduction of radar bombing equipment spurred this trend.

The Luftwaffe had sustained huge losses in bombers during simultaneous operations on the Russian and Mediterranean fronts. Its fighter force was nearly decimated. Available planes were grounded due to lack of fuel. As the year passed, Allied air forces found fewer and fewer fighters in the skies, so they targeted planes on the ground, trains, trucks, canal barges, and flak (antiaircraft artillery) towers. Such work did not appeal to the pilots, for low-flying planes were vulnerable to antiaircraft guns.

The Italian Macchi C.202 Folgore was actually equipped with a German engine. This marked a great improvement in performance over its predecessor—the Macchi C.200 Saetta fighter.

Women's Airforce Service Pilots

Pilots in Britain's Air Transport Auxiliary (ATA) shuttled planes back and forth among bases and delivered planes to the RAF from factories and foreign countries. Women pilots became part of the ATA in 1940. The work was dangerous: pilots were often shot at, and fuel could run perilously low in flights over the Atlantic.

Women pilots in the United States also joined the war effort. By 1941, American women were being accepted on a trial basis into the ATA. Some men protested vigorously, but with trained male pilots in short supply, the ATA had little choice.

American racing pilot Jackie Cochran served with the ATA before founding the U.S. Women's Airforce Service Pilots (WASPs) in 1942. Describing her experience in this effort, Cochran said: "We landed planes like the Hurricane and the Spitfire in fields where I wouldn't land my Lodestar today if I could avoid it."

In addition to ferrying Thunderbolts and making cargo runs, WASPs performed many other roles in aviation throughout the war. They worked on production lines, as mechanics on flight lines, as air traffic controllers, as flight trainers, and in radar stations.

The *Memphis Belle,* the first B-17 to complete its quota of 25 missions, was one of the most publicized aircraft of the war. When the celebrated bomber and crew returned triumphantly to the United States, a nationwide tour was undertaken to bolster sales of U.S. war bonds.

This rotating ball turret of the Boeing B-17 Flying Fortress *Memphis Belle,* which was attached to the underside of the aircraft, was also the only shelter for its operator.

Even at this late date, Allied area-bombing attacks on cities continued. Bomber Command attacked Berlin from November 1943 through March 1944, claiming that its four-engine Lancaster bombers could finish off the Germans' resistance and render the upcoming Allied invasion unnecessary. But 35 raids resulted in 1,047 bombers lost and 1,682 damaged—and no change in German resolve.

D-Day and Beyond Just prior to and after the successful, massive Allied landing in France on D-Day—June 6, 1944—most Allied air forces were diverted from air offensives (the heavy bombers only temporarily) to provide support for ground forces. The number of planes involved was impressive: the U.S. Eighth and Ninth Air Forces alone possessed 2,900 heavy bombers, 3,000 fighters, and 800 medium bombers.

The planes pushed forward, providing air cover for the Allied armies as they marched through France. The army slowed down only when weather prevented the advance of air forces and when the troops outstripped their lines of supply.

Air superiority enabled the Third Army, commanded by General George Patton, to dash swiftly toward Germany. So effective was the air cover that 20,000 German soldiers, attempting to turn Patton's flank, surrendered to the air force without ever engaging the Third Army. As Patton said of the air support for his armored divisions, it was "love at first sight."

The Bombing of Dresden Throughout the war, air forces had proved superb at tactical support of ground forces. However, the effectiveness of bombing cities and towns primarily by the RAF Bomber Command, often entailing a monumental destruction of lives and matériel, was sometimes questioned. The February 1945 fire-bombing of Dresden seemed a case in point: Why, just months before the war ended in Europe, did the Allies attack a civilian center? Dresden was reduced to ruins, and the flames took a week to die down.

The Curtiss P-40, of Flying Tigers fame, still captures the imagination of the public. Decades later, reconditioned P-40s hold great appeal at U.S. air shows.

The Pacific Theater After Midway

In April 1943 the Japanese lost their best military strategist, Admiral Isoroku Yamamoto. Because of the Navy's decoding expertise, a U.S. squadron of 14 P-38 Lightnings knew the exact moment he would pass by their ambush. His plane, along with another "Betty" bomber, went down in flames.

Island by island, the Americans moved closer to Japan, but each advance forward was costly. Americans attempted to bomb islands into submission before invading, but the Japanese, who hid in complex tunnel systems, emerged ready to fight as the U.S. Marines struggled across the razor-sharp coral reefs toward the beaches.

The Battle of the Philippines

The Battle of the Philippine Sea in June 1944 was proof of the growing superiority of American equipment and combat techniques. The Japanese lost 411 carrier planes and 3 carriers; only 50 American planes were shot down. The victory, and the capture of the Marianas, placed the Allies within reach of Japan.

At that point, General Curtis LeMay was brought in to orchestrate the bombing of Japan itself. The results of the initial raids using the new B-29 bomber had been disappointing. LeMay ordered the planes filled with incendiary bombs, stripped of guns, and flown low at 7,000 feet. Initially, the plan seemed foolhardy, but the results of this bombing raid were indisputable. Sixteen square miles of Tokyo were enveloped in the firestorm.

Japan's Mitsubishi Zero is shown with a Lockheed P-38 Lightning. The P-38 was the only U.S. fighter built before the war to remain in production at the end of hostilities.

U.S. Planes Take on the Zero

"Big Hog, this is Judah. Bogie zero-zero-zero, distance three-five, angels fifteen. Buster."

Through such radio messages, pilots communicated with each other in the air. (The message means: a single enemy aircraft has been spotted due north, 35 miles out, at an altitude of 15,000 feet; proceed at best speed.) Life was hard in the Pacific. Besides the threat of Japanese pilots, there were other trials: malaria, jungle rot between the toes, the fear of sharks. Downed pilots often were saved by islanders who, displaying a true wartime entrepreneurial spirit, sold them to Allied coast-watchers for sacks of rice.

In 1942, Japan's empire stretched from Indonesia to Southeast Asia to the Aleutian Islands of Alaska.

The Vought F4U Corsair, called the "Bent-winged Bird" by Americans, was the largest fighter to operate from aircraft carriers. The Japanese gave it a different nickname, which expresses their opinion of the Corsair: "Whistling Death."

The Lockheed P-38 Lightning The Allies had what they needed to win: planes that could keep up with the Zero, and plenty of them. The Lockheed P-38 Lightning could deliver a bomb load equal to that of a medium bomber, and it had amazing climbing ability. Later models of this plane used by U.S. forces had turbo-superchargers and were fearsome opponents.

The Grumman F6F Hellcat The Grumman F6F Hellcat was produced prolifically for use by Pacific fighter squadrons. From the day of its appearance, the Allies were destined to win the war in the Pacific. By 1943 it was the plane most frequently flown from aircraft carriers, destroying about 5,000 enemy aircraft. It often worked in coordination with the Corsair; the planes together were known as the Terrible Twins.

The Vought F4U Corsair Finally, there was the Vought F4U Corsair, with its distinctive inverted gull wings. At first the plane was considered unfit for carrier operation. However, the Corsair was eventually used on carriers and also operated effectively from land. Charles Lindbergh, a technical adviser and pilot in the Pacific, expressed the power its pilots felt: "Guns charged and ring sights glowing, our four Corsairs float like hawks . . . over the jungle hills . . ."

The B-29 Superfortress Boeing's B-29 began service in the Pacific in June 1944. Unlike the B-17 and B-24 bombers, its interior was pressurized, increasing crew comfort and efficiency at higher altitudes. Armed with machine guns and cannon, the B-29 could deliver a bomb load of 20,000 pounds.

Kamikazes

In a desperate effort to regain the advantage, Japan launched kamikaze ("divine wind") attacks against U.S. aircraft carriers. Flying a plane loaded with explosives, the kamikaze pilot had one goal: to strike the enemy ship before enemy fire stopped him. It was a deadly tactic: attacks by kamikazes accounted for heavy U.S. casualties, and for half of all U.S. Navy ships damaged and one-fifth of all those sunk in the latter half of the war. It was also deadly for the pilot—there was no surviving a kamikaze flight.

A pilot preparing for a kamikaze run shaved off his hair and bound his head with a Japanese flag. Then the pilot cut off a part of his body, such as a lock of hair or one of his fingers, which would be cremated and the ashes sent to his family shrine. In the air, the pilot would use every ounce of skill to hit the ship before his plane was destroyed. A Japanese pilot described how a kamikaze fought his last battle:

> He screams unscathed through the barrage, leveling inside the flak umbrella near the water. A hit! He's struck a destroyer right at the water line. A bellowing explosion, then another and another. It's good! It's good! The ship is in its death throes. It can't stay afloat—water plunging over the bow, stifling it. It upends and is gone.

The Atomic Bomb

By the middle of 1945, U.S. Navy carrier aircraft had shot down 6,484 Japanese planes and destroyed another 6,000 on the ground. They had also sunk 700,000 tons of enemy combat vessels and 1.3 million tons of merchant shipping. Still, Japan had not surrendered.

Hiroshima President Harry Truman's decision to end the war without any more loss of American life led to the flight of the Boeing B-29 bomber *Enola Gay* (named for the mother of the pilot, Colonel Paul Tibbets, Jr.) over Hiroshima.

Boeing B-29 bombers (right) first entered military service in June 1944. The deadly cargo of the B-29 *Enola Gay*—the uranium bomb Little Boy (below)—changed the nature of modern warfare and ushered in the nuclear age.

Truman ordered that a newly developed atomic bomb be dropped on a Japanese city. On August 6, 1945, at 9:12 A.M., the *Enola Gay* reached its Initial Point over Hiroshima, marking the start of the bomb run. A radio tone signaled the automatic drop of a uranium bomb in 15 seconds. The pneumatically operated bomb bay doors opened, and Little Boy #1 fell out. The plane jumped up, five tons lighter, and the radio tone stopped.

After a sharp diving turn to the right, the *Enola Gay* moved swiftly away. The crew's eyes closed instinctively with the flash—the instantaneous generation of 100 million degrees of heat.

A spectacular mushroom cloud formed. One crewman described it as a bubbling mass of purple-gray smoke with a red core; to another, it looked like lava or molasses covering the city and flowing into the foothills.

Then came the shock wave. One crewman said it felt like a giant hitting the plane with a telegraph pole. On the ground, some 75,000 people had just died.

Nagasaki Even so, Japan did not surrender. Another eight days passed before the Japanese bowed. By that time, Nagasaki had become the target of a second B-29, this one carrying a plutonium bomb. The bomb killed an estimated 35,000 people and forced Japan's surrender.

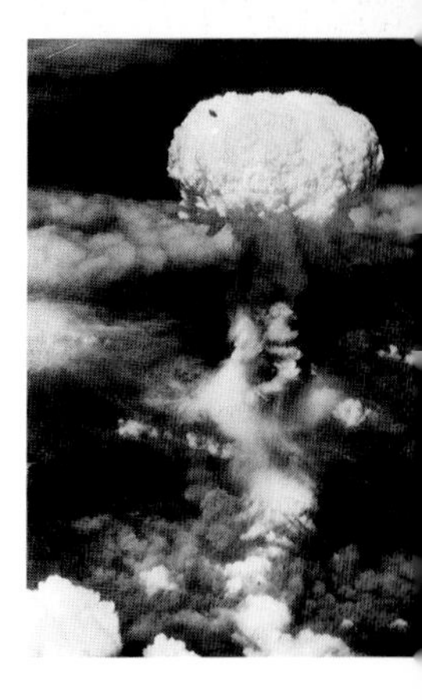

Under a program called the Manhattan Project, the United States became the first country to produce atomic bombs. The second atomic bomb, known as Fat Man, was dropped on Nagasaki. It was a plutonium bomb, more powerful than Little Boy.

The Jet Age

Toward the end of World War II, the Allies and the Germans were building the first jets—airplanes that could fly at unheard-of speeds.

Jet technology brought swift changes to military craft and civilian airliners. New high-speed fighter tactics were deployed in the Korean and Vietnam wars. Jet engines revolutionized passenger transport, spawning jumbo jets and a supersonic transport.

Charles "Chuck" Yeager poses with the Bell X-1, *Glamorous Glennis,* which he piloted past the sound barrier in 1947.

Jet Fighters, 1940–1950

Research into turbojet engines (driven by a powerful discharge of exhaust rather than by pistons) and design for jet aircraft were costly, and the aircraft industry hesitated to risk the investment without a known market for passenger jets. The interest of air forces in jet fighters and bombers created a market and proved the workability of jets before the idea of jet-powered passenger craft became feasible.

The Early Jets of World War II

Research in Germany Nazi leaders respected the potential of air power and provided financial support for jet research, generating several breakthroughs.

German designers were the first to raise the issue of *compressibility*—the high drag and control problems encountered at high speed. Starting in 1935, German engineers worked with swept-back designs to reduce drag. In August 1939, German designer Hans von Ohain saw his Heinkel He S 3b gas turbine successfully power the first flight of the Heinkel He 178 jet aircraft.

The Me 262 Prototypes of the world's first operational jet fighter were being tested as early as 1941. The airplane was unquestionably a record-beater, but Hitler ordered the Messerschmitt Me 262 to be reworked as a bomber. This switch in design, together with engine problems, delayed the plane's arrival on the front lines until 1944; it was eventually released as a fighter but too late to have a decisive role in World War II.

One Allied pilot marveled at the plane's beauty and elegance: "It looked superb with its triangular fuselage like a shark's head, its tiny arrow-shaped wings, its two long turbines, its grey camouflage spotted with green and yellow."

With a speed of 540 miles per hour at 20,000 feet, the Me 262 cut through the Allied fighters protecting B-17 bombers. As late as March 1945, when Allied fighters outnumbered German aircraft in the skies, the Me 262 was still causing the Americans substantial bomber losses.

The Messerschmitt Me 262 was first flown in July 1942—the first operational turbojet-powered airplane. Both a fighter and a bomber version were built.

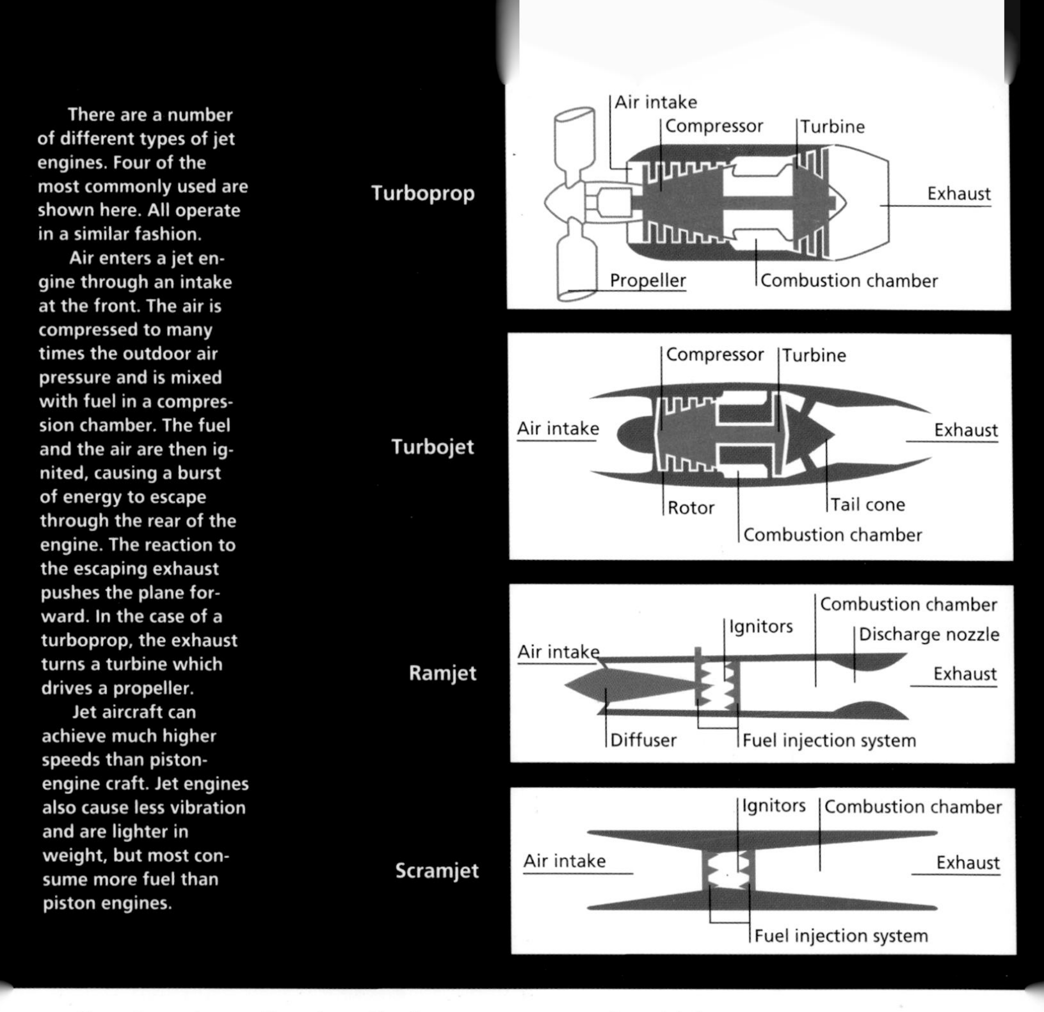

There are a number of different types of jet engines. Four of the most commonly used are shown here. All operate in a similar fashion.

Air enters a jet engine through an intake at the front. The air is compressed to many times the outdoor air pressure and is mixed with fuel in a compression chamber. The fuel and the air are then ignited, causing a burst of energy to escape through the rear of the engine. The reaction to the escaping exhaust pushes the plane forward. In the case of a turboprop, the exhaust turns a turbine which drives a propeller.

Jet aircraft can achieve much higher speeds than piston-engine craft. Jet engines also cause less vibration and are lighter in weight, but most consume more fuel than piston engines.

One American pilot described an engagement in which two Me 262s attacked a bomber fleet in full view of the 48 P-47 Thunderbolts escorting them. The jets came in fast—the Thunderbolts had no chance to intercept them before the bombers were struck by their fire. All the P-47s dived on the two Messerschmitts, but "even in a near-vertical dive at full power, and with our altitude advantage, we didn't even come close. Their speed in level flight was absolutely amazing to us."

Due to Germany's lack of fuel, many of the Me 262s produced at the end of the war never left the ground. Parked at airfields, they were destroyed by the Germans as the Allied tanks approached.

The Gloster Meteor The first British jet fighter, the Gloster Meteor, powered by a jet engine designed by Frank Whittle, entered service almost simultaneously with the Me 262 in 1944, but did not have the same impact. The Messerschmitt could fly more than 100 miles per hour faster than the Meteor. Both planes had a fairly conventional design. The difference in speed resulted from the greater thrust developed by the Me 262's jet engine, and the craft's superior aerodynamics. Still, the Meteor retains the distinction of being the only Allied jet to see action during World War II.

U.S. Fighter Jets

When war broke out in Europe in 1939, the British and the Germans were developing aircraft for military service. In the United States, fighter plane design and production were at an earlier stage of development, and the U.S. Army Air Corps commissioned Bell Aircraft to design the first jet.

The Bell P-59 Airacomet In November 1941, the first P-59 flew. Powered by British-designed engines produced by General Electric, the P-59 was a remarkable achievement, equaling the performance of the early Meteors. Still, the plane fell short of expectations. Classified as a fighter-trainer, the P-59 had a limited production run—only 66 were built.

The Lockheed F-80 Shooting Star The Air Corps then turned to the Lockheed Aircraft Company to develop a better jet fighter. The designers started work on May 17, 1943, and built a prototype for the F-80 Shooting Star in just 139 days.

Like the other early jet fighters, the F-80 was an example of a rather conventional design equipped with a turbojet engine. In early test flights, engine problems were responsible for several fatal crashes.

By 1947, the F-80 had been greatly improved, and Shooting Stars reached fighter squadrons by the beginning of the Korean War. In fact, the world's first all-jet air combat engagement was fought between an F-80 and a Russian MiG-15 on November 8, 1950. The American craft was victorious, but the MiG was clearly the superior fighting machine. After the war, Shooting Stars were relegated to service as support planes or trainers.

The Lockheed XP-80 was the prototype for the F-80 Shooting Star. It featured low, cantilever wings and a turbojet engine enclosed within the rear fuselage. The airplane set a world speed record of 623 miles per hour in 1947.

Wings for Jet Airplanes

As a jet airplane approaches the speed of sound, air molecules in front of it cannot move out of its path; instead, pressure waves build up and buffet the aircraft.

To counter this effect, designers have created sleeker airframes with needle noses, fuselages with nipped-in waists, and thinner wings with sharper edges to slice through the compacted air.

Swept-back wings and delta wings are two particular features that allow jets to fly at these speeds. Most jet airliners are built with swept-back wings, which hit the air at an angle and allow the airplane to reach higher speeds before pressure waves build up. Supersonic transports and many fighter jets are built with delta wings, whose superthin, broad expanses provide extra lift and low drag to optimize speed.

The Airbus A319 (top) clearly shows the swept-back wings common on civilian jets. The delta wings of Britain's Fairey Delta (bottom) allow the speed and maneuverability that a jet fighter needs.

Jets in the Korean Conflict, 1950–1953

Whatever the limitations of the Shooting Star, at least it was jet-powered. When the Korean conflict began in 1950, the U.S. Army Air Corps chose to take P-51 Mustangs and F4U Corsairs out of mothballs. Pilots agreed they were great old planes, but war was no place for nostalgia. Five years after World War II, any fighter that lacked jet power was obsolete. Pilots required a jet's speed to win in an engagement.

The U.S. Air Force, formed in 1947, first fought as an individual arm of the military in the Korean War.

North American's F-86 Sabre, the first jet fighter built with swept-back wings in the United States, distinguished itself in the Korean War. A heavier version of the original F-86, the F-86D Sabre, (above) was produced in great numbers, but was a disappointing performer.

The F-86 Sabre Versus the MiG-15 The U.S. Air Force fielded one fighter in Korea that was more than equal to its competition: the North American F-86 Sabre. Facing off against it was the Russian-built MiG-15.

The Sabre and the MiG were considered roughly equivalent aircraft. Some of the technology for each craft was skimmed from other countries' research: the Russians bought the MiG-15's Rolls-Royce engine from the British, and the Sabre's swept-back wings were a bonus of the German research data acquired by the United States after World War II. Also, a designer at the National Advisory Committee for Aeronautics (NACA) had discovered the wing's drag-reducing advantages.

Fighter engagements sped up dramatically with these craft: approaching each other, the planes could close up 10 miles in 30 seconds. At that speed, as one pilot noted, everything happened fast: "Suddenly contrails broke out straight ahead and high above us, followed by bursts of sunlight flashing off MiG canopies and fuselages . . . Red-nosed, stubby aircraft with big red stars flashed by . . ."

MiG Alley Korea's Yalu River valley became known as MiG Alley. UN forces were not allowed to chase the MiGs north of this line, so the North Korean and Chinese air forces massed fighters just behind it before flying south to strike their targets. In the brilliant blue of MiG Alley, stretching from the ocean to the sky, the combatants grappled.

The Russian design team of Mikoyan and Gurevich (MiG) built the prototype for the MiG-15 in 1947. More than 1,000 of these lightweight and agile craft were purchased by China and North Korea.

A round disc called a radome, which houses a radar antenna, makes an airborne warning and control systems (AWACS) airplane easy to recognize. Crew members on board an AWACS use electronic surveillance equipment to gather intelligence.

Though often outnumbered, the Sabres dominated the skies, with a victory ratio of more than 10 to 1. But though the F-86 was a better gun platform than the MiG, most of the difference lay in the pilots, not the machines. The North Korean pilots were green, whereas many of the Americans were experienced World War II veterans. The Air Corps was impressed with the MiG and offered an award of $100,000 for the first pilot to deliver a MiG-15 across the lines.

New Tactics for Bomber Escorts Although pilots on both sides respected enemy planes, neither the Sabre nor the MiG worked well as a bomber escort. The Sabres shot down a lot of MiGs, but the MiGs destroyed a lot of bombers.

The problem was straightforward: the speed of jet fighters required a change in tactics. Air battles to protect bombers could no longer occur in the middle of the bomber formation. Instead, the enemy had to be fought at a minimum of one hundred miles ahead of the bomber's target. Pilots needed radar surveillance and guidance either from ships or from airborne warning and control systems (AWACS), to prepare that far in advance. Neither option was available in Korea, so the Americans stopped daylight bomber raids over North Korea.

G Forces

In some maneuvers, jet fighter pilots face sudden accelerations, abrupt slowdowns, and sharp turns, which place the human body under great stress from G forces (a measure of the force of gravity). The normal pull of gravity at sea level is equivalent to 1 G. Higher G forces make it hard for the blood to circulate. At 2 Gs, peripheral vision starts to fade; at 6 Gs, the arms can't be lifted and the pilot is in danger of blacking out. Fighter pilots now wear G suits to provide some protection against high G forces, and computer controls prevent maneuvers that endanger the pilot.

Supersonic Fighters, 1950–1975

In the five years after World War II, jets pushed piston-engined planes out of fighter squadrons. In another five years, supersonic fighters were ousting the earlier jets. (Despite the pace of change from the 1950s onward, the supersonic F-4 Phantom II has remained in front-line service for more than 25 years.)

The technology of supersonic fighters vastly changed the role of the pilots—sometimes to their disadvantage. The F-4 Phantom IIs, for example, went into Vietnam armed with radar-guided and heat-seeking missiles and no machine guns or cannon. If the sophisticated weapons systems failed, as they sometimes did, the pilot was left defenseless. Soon, American fighter planes carried both missiles and guns. The navy founded the Navy Fighter Weapons School (Top Gun School) in San Diego, California, to teach pilots the new art of supersonic air combat with an array of weapons.

More than 5,000 McDonnell Douglas F-4 Phantom II aircraft, shown here on active duty, were manufactured before production was halted in 1979. The U.S. Air Force's F-4s were custom-designed for air-to-air missile combat and were used by both navy and air force crews.

The Convair F-102 Delta Dagger Studying German research data after World War II, Convair engineers became convinced that a delta-wing design would be ideal for supersonic craft. The delta wing would be more stable and maneuverable at high speeds, and its greater strength would permit the use of thinner, more rigid wings.

When F-102's prototype, the XF-92A, flew on September 18, 1948, it became the world's first delta-wing airplane. Test flights were generally successful, but the plane could not exceed the speed of sound. To remedy that problem, the engineers employed a theory, called the area rule, developed at the NACA Langley Aeronautical Laboratory. It posits that the fuselage should be slimmer where the wings attach to it (the "bumble bee" shape).

The engineers thinned the midsection of the plane and lengthened and widened the rear section. The modified prototype flew on December 20, 1954, exceeding Mach 1. Nicknamed "the Deuce," the F-102 Delta Dagger saw service in fighter squadrons throughout the rest of the 1950s.

The Fairey Delta The Fairey Delta 2 was the first British supersonic airplane. Powered by a Rolls-Royce Avon engine, it achieved supersonic status less than a year after the F-102. In March 1956 the plane set a new world airspeed record of 1,132 miles per hour, at 38,000 feet.

The Fairey Delta's "droop snoot" nose, which allowed the entire nose and cockpit to hinge downward by 10 degrees, gave the pilot better forward vision during takeoff and landing. The same design later reappeared on the Concorde and Tu-144 supersonic airliners.

Convair F-106 Delta Dart In the 1960s and 1970s, the F-106 served as the backbone of the U.S. defense against supersonic enemy bombers. Often described as the best interceptor ever built, the plane stayed on the front line as long as the F-4.

The delta-wing Convair F-106 Delta Dart had a top speed of 1,500 miles per hour. Its sophisticated electronic system was regularly updated, enabling the F-106 to serve the U.S. Air Force as an interceptor from 1958 to 1988.

Breaking the Sound Barrier

Built of high-strength aluminum, the Bell X-1 was equipped with a 6,000-pound-thrust rocket engine.

On October 14, 1947, a human being traveled faster than the speed of sound for the first time. Charles "Chuck" Yeager flew the Bell X-1 research craft, called the *Glamorous Glennis,* past the sound barrier into supersonic flight. The rocket-driven, bullet-shaped plane was named for Yeager's wife.

To conserve fuel, a B-29 "mother airplane" carried the X-1 in its bomb bay to a height of 12,000 feet. There Yeager, wearing a flight suit, leather jacket, and oxygen mask, climbed down from the B-29 into the X-1's cockpit. He was in considerable pain from two cracked ribs sustained in a horseback-riding accident, but he insisted on making the flight anyway.

The X-1 was launched in midair. As it approached Mach .96, the aircraft hit the turbulent shock waves that Yeager had experienced in earlier test flights. At 43,000 feet, Yeager pushed the plane past the sound barrier to Mach 1.05 (700 miles per hour). He found that the buffeting stopped—the X-1 was easy to control. The ground crew heard a new sound: a sonic boom.

After a few moments of supersonic flight, Yeager cut the rocket engines, steadied the X-1 through the sound barrier's shock waves, and landed the craft safely at the Muroc (California) air base.

First tested in 1955, the MiG-21 became one of the most widely used fighters in the world; more than 20 variations have been produced. Typically armed with a twin-barrel gun, it is equipped with pylons beneath the wings for weapons or fuel tanks.

The F-4 Phantom II and the MiG-21 The Vietnam War's version of the Sabre–MiG-15 dogfight in Korea was the F-4 Phantom II facing off against the supersonic MiG-21.

The MiG-21 was a delta-winged, Mach-2 aircraft, armed with a multibarrel cannon and air-to-air (to hit other aircraft) and air-to-surface (to strike ground targets) missiles. Highly maneuverable, it was an extremely dangerous opponent.

The F-4, on the other hand, was the first fighter able to identify, intercept, and destroy targets within range of its radar without assistance from ground control. This Mach-2 aircraft, which carried 16,000 pounds of weapons, filled more roles in Vietnam than any other plane—air support for ground troops, air-to-air combat, even bombing. It excelled as a high-speed, high-level interceptor but performed equally well at lower speeds at low altitude. The Phantom broke scores of records, among them the world speed record—at 1,606 miles per hour, on November 22, 1961. It destroyed 145 MiGs in Vietnam.

A legend in its own time, the F-4 spawned various models, which served in the air forces of twelve nations. By the time production slowed in 1978, 5,000 F-4s had entered service.

Mach Numbers

Named for Austrian physicist Ernst Mach, Mach numbers describe the speed of an aircraft in relation to the speed of sound. Mach 1 is the speed of sound—760 miles per hour at sea level. Sound travels faster in warmer air; therefore, in the cold air at 40,000 feet, the speed of sound, Mach 1, is only 660 miles per hour. Mach 2 is twice the speed of sound, Mach 3 three times the speed of sound, and so on.

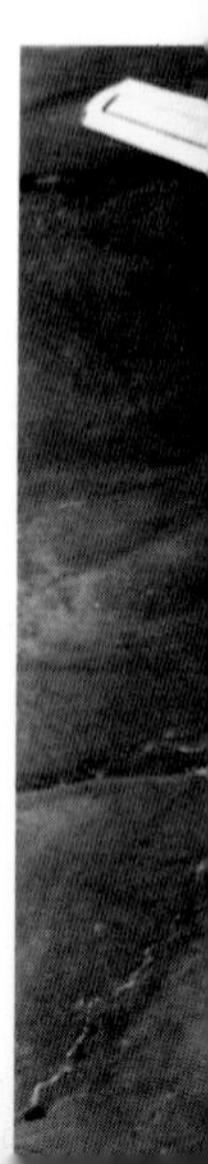

Jet Bombers and Transports, 1945–1975

Jet Bombers

Long-range bombing of military and industrial centers has been a key military strategy since the early days of World War II. Thus, established bomber designers, such as Boeing, eagerly considered how the swept-wing jet design could be applied to the big planes.

Two Boeing models, the B-47 and the B-52, stand out from the crowd. Experience with these huge planes led directly to the development of passenger jets and military transport aircraft.

The B-47 The development program for the B-47 was set at $10 million—a tiny amount by today's standards. Boeing risked millions to create the bomber, with its narrow, swept-back wings, underslung pods for jet engines, bicycle landing gear, and drag parachutes to slow the plane on landing.

In 1947, the plane emerged as the most important jet bomber of the time. There were no other state-of-the-art bombers—a cold war advantage that made the B-47 the plane to fly for up-and-coming military pilots. The U.S. Strategic Air Command (SAC) relied heavily on its force of 1,650 B-47s throughout the 1950s.

The 100-ton B-47 bomber was powered by six turbojet engines set in pods and suspended on pylons from the swept-back wings. The B-47 could carry a 20,000-pound bomb load and reach a maximum speed of 606 miles per hour.

A parachute is used to slow down the landing of a B-52. In compliance with the Strategic Arms Reduction Treaty, the U.S. Air Force began destroying 350 B-52 airplanes in late 1993. After Russian officials inspect the destruction of the aircraft by satellite photo, the B-52s, built at a cost of $64 million, are then sold as scrap for about 16 cents per pound.

Still, almost immediately, SAC began planning for the B-47's replacement. General Curtis LeMay, the head of SAC, wanted an airplane that was big enough to take on a variety of functions as needs arose. The B-52 fit that role well, with an airframe and equipment vastly more sophisticated than those of the B-47.

The B-52 The B-52 prototype flew on April 15, 1952. At that time, it was slated to be a high-altitude nuclear bomber, capable of single-plane missions over the Soviet Union. Twenty years later, in Vietnam, the versatile plane filled an entirely different role as a low-flying platform for dropping conventional bombs.

Because of its all-weather capability, the B-52 was well suited for use in Vietnam, particularly as hazardous flying conditions grounded most other military aircraft during the monsoon season. At the same time, bomber crews were limited by vacillating policies on the part of the U.S. government, sometimes leading to heavy bombing of South Vietnam, the ally, rather than enemy territory in North Vietnam.

When the North Vietnamese, whose strong air defenses had been underestimated by U.S. analysts, stalled at the peace talks, a U.S. operation named Linebacker II began saturation bombing. In a little over a week, B-52s dropped 15,000 tons of bombs on Hanoi and Haiphong, both heavily defended targets. The bomb strikes overwhelmed the North Vietnamese, who returned to the negotiating table.

The B-52 was so well established as the backbone of SAC that officers were comfortable making fun of it, calling it the BUFF (big ugly fat fella). It would still be in action twenty years after Vietnam in the Persian Gulf War.

Supersonic Bombers

The Convair B-58 By the 1950s, bombers such as the Convair B-58 Hustler were ready for supersonic flight. A completely original design, the B-58 carried its bomb load in an external pod formed from panels of bonded honeycomb sandwich skin to resist heat. Like the B-47 and the B-52, the Hustler used a parachute to slow it down on landing.

The B-58 had no equal anywhere in the world from 1956 to 1970. Its service life was cut short, however, by high operating costs and a high accident rate. Of the 104 aircraft built, 25 were lost to crashes.

Some other supersonic bombers were killed on the drawing board because of changes in government policy. Unpiloted intercontinental missiles had taken over much of their role. With missiles, neither air crews nor planes are placed in danger over enemy territory.

Taking the pilot out of the bombing raid has its disadvantages, though. Missiles cannot be called back, nor their mission changed, once they have been launched.

Cargo Planes

Bombs are not the only heavy load air forces carry. Though the big planes that transport troops, equipment, food, and medical supplies around the world are not as flashy as the fighters, they play a vital role.

The Vietnam War demonstrated the reliance of troops on cargo planes, which carried multibattalion forces into battle areas and kept them supplied with fuel, food, and ammunition. Day after day, Lockheed C-130s landed in the midst of shelling, dropped off supplies, and took on wounded. In 1967, at Dak To, C-130s landed 5,000 tons of supplies on a shell-damaged runway.

The Berlin Airlift The Berlin airlift is a famous example of using air transport as a lifeline. The Soviet Union, in an effort to pressure the Allies out of West Berlin, blockaded the city in 1948. When the U.S. military governor asked the air force if it could carry coal in, the answer was, "Sir, the air force can deliver anything."

Cargo planes carry anything anywhere—even cows.

The versatile Lockheed C-130, known to its crews as "Herky-bird," first flew in late 1954. A veteran of the Korean, Vietnam, and Persian Gulf wars, C-130s are still in service for tasks such as air-dropping cargo to otherwise inaccessible areas and refueling in midair.

Using Douglas C-54s and other cargo planes, the air crews made good on that promise. At each of the two airports used for the airlift, a C-54 landed every three minutes. Off-loading was frenetic: one 12-person German crew managed to unload 12,500 pounds of coal in 5 minutes—a record. Between June and September, the transports delivered 2,325,000 tons of food, fuel, and supplies. The crews flew 189,963 flights covering a total of 92,061,863 miles.

Lockheed's Starlifter and Galaxy Later developments in transport planes included the Lockheed C-141A Starlifter, introduced in 1965, and the Lockheed C-5A Galaxy, introduced in 1969. Crews particularly enjoyed the C-141A, which was a reliable, comfortable plane to fly.

After use as gunships in Vietnam, production of C-130s has continued for military purposes. C-130s and C-5As also fly humanitarian missions. By the early 1990s, C-130 production had reached the 2,000 mark.

Jet Airliners

Military contracts covered the £22 million needed to develop the Rolls-Royce Avon engine, as well as the $150 million needed for the Pratt & Whitney J-57, the first two engines efficient enough to make jet airliners safe and profitable. Commercial aviation has benefited greatly from these technological advancements.

More than one million passengers flew the Atlantic in jet airliners in 1958—the first time that planes carried more people on that route than steamships did. The airplane was outstripping ships and trains as the dominant passenger transport. Jet traffic across the Atlantic topped 6 million passengers by 1968, and 30 million passengers by 1992.

In 1958, Boeing grabbed the lion's share of the jet airliner market with its 707. McDonnell Douglas Aircraft lagged a step behind with the DC-8. These two planes became the most successful money-makers in air transport history. The public's en-

thusiasm for flying encouraged the aircraft companies to proceed with wide-body, large-capacity jumbo jets, carrying three hundred to five hundred passengers each, and led to development of the supersonic transport (SST).

The gleaming Lockheed Constellation was the star of postwar passenger travel before the arrival of jet airliners. Pan Am, TWA, and American flew the luxurious, four-engine "Connie" on intercontinental routes. Introduced in 1943, it endured through the 1950s.

The de Havilland Comet

The story of the world's first jet airliner is one of meticulous development marred by a single, tragic flaw. Five years before the Boeing 707 rolled out of the factory, the de Havilland Comet went into service with the British Overseas Airways Corporation (BOAC), now British Airways. "Whether we like it or not," the editor of *American Aviation Magazine* had to admit, "the British are giving the U.S. a drubbing in jet transport."

De Havilland Takes Up the Challenge That drubbing started taking shape at the end of World War II, when Geoffrey de Havilland was approached by a committee of airline executives, manufacturers' representatives, and aviation experts. They were looking for a British aircraft that could challenge the DC-3's clear lead in the passenger transport business. The committee had settled on a jet aircraft as the best solution—though it would be difficult to achieve.

As designer of some of Britain's best warplanes, de Havilland had already begun to work with jet engines during the war. He knew that any aircraft he built would have to fly at altitudes of 35,000 feet and higher. Jet engines of the time consumed three to four times as much fuel at 10,000 feet as at 30,000 feet. Flying high would solve the fuel problem.

The Comet 4 entered service in October 1958. It was flown by commercial airlines, and its military version, the Nimrod, saw service with the RAF. The Comet 4 was a sound craft, built to overcome the first Comet's structural weakness.

It would, however, create another problem: pressurization. The aircraft's cabin would have to be airtight and pumped full of air so that the passengers could breathe without oxygen masks. Pressurized aircraft had appeared before the war, but none had flown very high. In the thin air at 35,000 feet, the difference between the inside and outside pressures would be greater than that ever encountered by a passenger aircraft, creating immense structural stress.

Design and Testing After considering several approaches, the design team settled on a craft with wings swept back by 20 degrees. To minimize weight, a light aluminum skin, only .028 of an inch thick, was attached to the plane's frame.

The team knew they were facing uncharted territory and therefore tested the Comet as no plane had been tested before. They decided that the Comet would have to withstand internal pressures of more than 20 pounds per square inch, though regulations demanded only 16.5 pounds. The wingtips were repeatedly raised and depressed by 3 feet off their normal positions, and the landing gear was extended and retracted countless times.

Jet Passenger Service Begins From the instant of its unveiling on May 2, 1952, the Comet caused a public sensation. Passengers felt comfortable in the pressurized cabin 8 miles up in the stratosphere. Vibration-free jet engines allowed them to sleep—an unheard-of luxury.

With four 4,500-pound-thrust turbojet engines, the Comet was the fastest commercial transport in the air. Its range of 1,750 miles was twice the distance that U.S. experts thought possible. And it was profitable. In its first year, the airliner carried 28,000 passengers a total of 104.6 million miles.

Then, in January 1954, a Comet disappeared over the Mediterranean. BOAC's fleet of seven airliners, those of Air France, and many in the service of other airlines were immediately grounded for modification, though the engineers could only guess at what needed to be done. The planes started flying again as salvage operations retrieved jagged pieces of the Comet from 100 square miles of ocean floor.

On April 8, two weeks after service had resumed, another plane flew into oblivion. The only mark of its passing was wreckage bobbing on the sea near Stromboli.

The Comet's Fatal Weakness A new testing method pinpointed the problem. Structural engineers built a watertight tank around the fuselage of one of the grounded Comets, with the wings protruding from holes on either side. Ballast took the place of passengers, and the cabin was flooded. By repeatedly raising and lowering the water pressure inside the cabin, the analysts simulated a lifetime of cabin pressurization cycles.

What they found was horrifying. After being subjected to the equivalent of nine thousand flying hours, the flooded Comet's fuselage split open. In flight, such a rupture would have caused the cabin to explode like a bomb. All evidence from the remnants pointed to the same conclusion: explosive decompression.

Permanently grounded, the Comet suffered the fate of a pioneer. Its design team had identified and solved many issues associated with high-level flight, but they had missed one key factor. When the Comet climbed quickly to altitude and descended quickly to land, rapid changes in pressure flexed the plane in a way that had not been anticipated. Metal fatigue in the corner of a window was the Comet's fatal flaw.

The Comet's four jet engines, encased in the low, swept-back wings, powered the craft to a maximum speed of 490 miles per hour. Flying 8 miles above the earth, it avoided the turbulence usually associated with air travel. Yet its graceful profile hid a tragic flaw.

Cockpit of Boeing 747-400

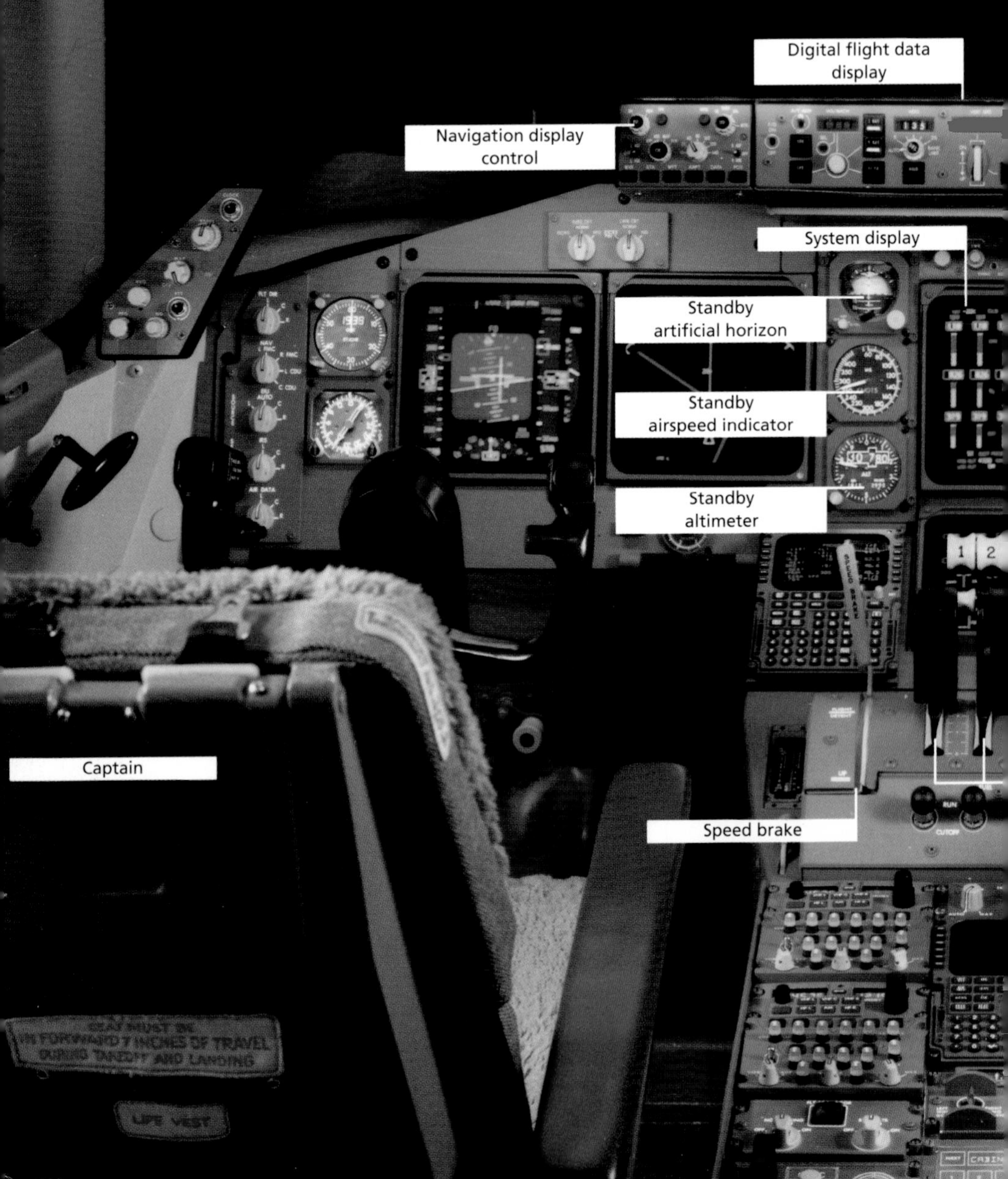

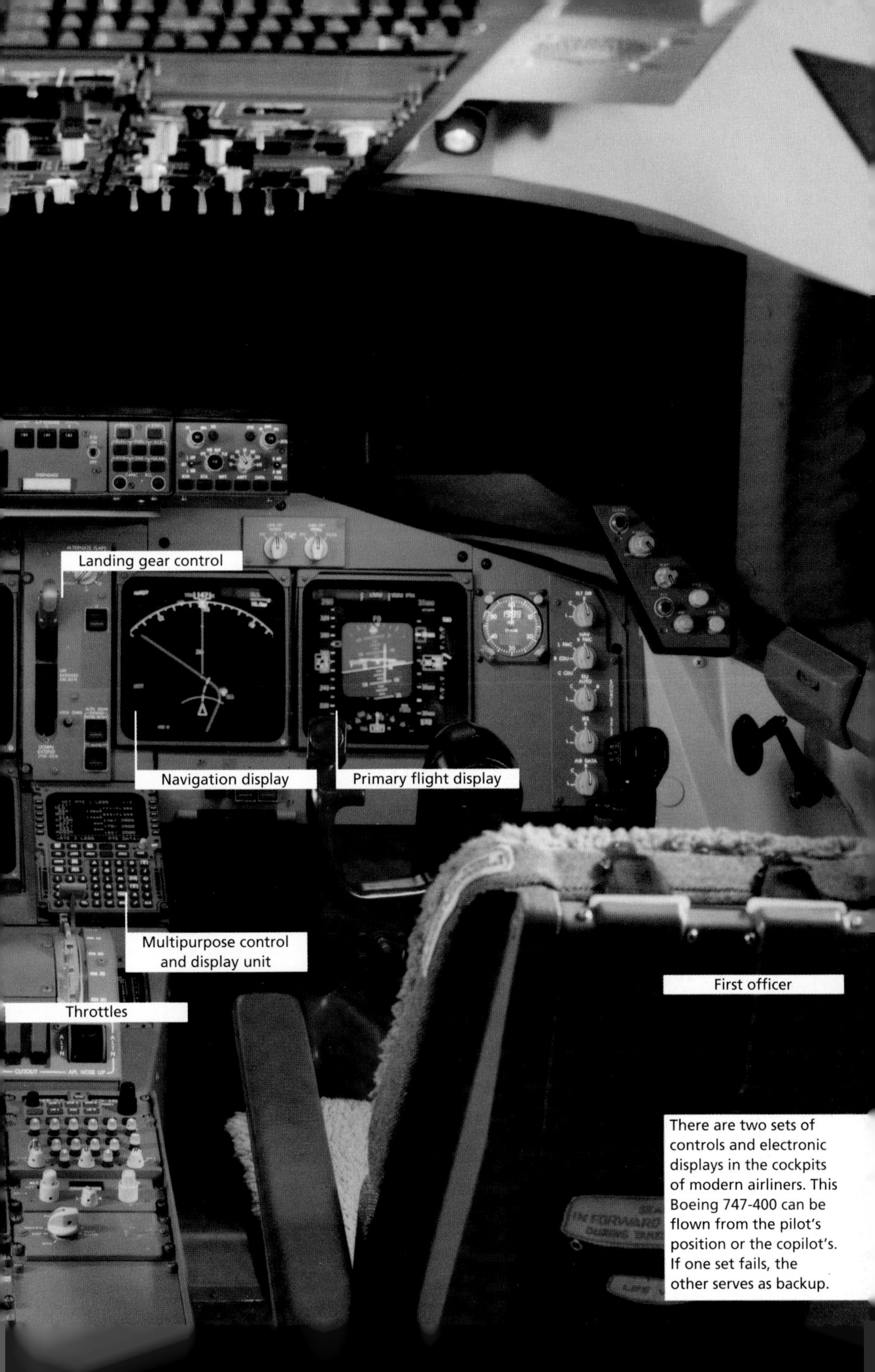

There are two sets of controls and electronic displays in the cockpits of modern airliners. This Boeing 747-400 can be flown from the pilot's position or the copilot's. If one set fails, the other serves as backup.

The Boeing 707

The next jet airliner to appear was the Boeing 707. The first 707 prototype, the 367-80 (or "Dash Eighty"), rolled out of a Seattle factory one month after the Comet's final crash.

Boeing undertook the development of the 707 before a single customer signed on the dotted line. It was certainly a calculated risk, but not a blind leap of faith. The Comet had proved that the public was ready to accept jet travel. Boeing also was planning to use the same airframe for both the 707 and a new jet tanker, the KC-135.

Design and Testing Boeing's experience in building bombers and transports probably influenced one of their wisest design decisions for the 707. Months before the Comet puzzle was solved, Boeing engineers specified that the aluminum skin of the 707 would be 4.5 times thicker than the Comet's. Inside the skin, they welded "tear stoppers," made of titanium, to block the spread of a metal-fatigue crack.

The design team tested the 707 in a way no aircraft had been tested before. First, the team built a wooden prototype, and later a flyable version, the Dash Eighty. Sensitive instruments and automatic cameras were attached to the Dash Eighty to gather data from a vast array of tests. The result was an astonishingly reliable craft, given its size—the 707 weighed 115,000 pounds empty.

Boeing reassured nervous potential customers with a sales movie, "Operation Guillotine," which opened with a shot of an airliner that looked suspiciously like the Comet. Two steel blades mounted above the fuselage dropped in slow motion. Where they pierced the plane, the metal skin split, the fuselage burst, and seats, dummies, and overhead bins spewed out.

The Boeing 707, first flown in 1954, made passenger jet airliners an unquestionable commercial success. Constantly improved and modified over its long life, the 707 continues to perform well, both in passenger service and in a variety of roles in the military.

In the next scene, five blades bit into the skin of a 707. The tense audience gasped, then relaxed. Little puffs of air escaped, but no explosion occurred. Obviously, if a tear occurred in a 707, the pilot would have time to descend to a safe altitude.

Boeing's first customer was an old friend: the U.S. Air Force. It requested a fleet of 29 jet tankers built from the same airframe as the 707. That order, in October 1954, was a wake-up call to Douglas Aircraft.

The Douglas DC-8

Some months later, in mid-1955, Douglas announced plans for a new jet aircraft, the DC-8. Some work had already been done on a DC-8 design, but Douglas had been diverted by customers clamoring for a new piston-engine aircraft. But Pan Am executives wanted to talk about jets when they met with Douglas officials the same year. Pan Am had been banking on the now-defunct Comet from Britain. Now the executives wanted to spur competition between Boeing and Douglas.

Design and Testing Douglas's designers identified the 707's weaknesses and used them as a starting point for their plane. The DC-8 would have a wider cabin and a greater range. Douglas's sales force urged customers to wait a year for a better product. The DC-8 design also was exhaustively tested.

By October 1955, Douglas pulled ahead of Boeing in the jet race. On that day, Pan Am announced an order for 25 DC-8s and 20 707s. The DC-8 had beaten Boeing's record-setting 707 by five orders. Following fast in Pan Am's footsteps, United Air Lines signed a contract for 30 DC-8s, because of the plane's greater cabin width, which added a passenger seat to each row.

The Douglas DC-8 has had a long life as both a passenger transport and a cargo airplane.

The Learjet

Boeing Fights Back At first, Boeing refused requests to widen the 707. But the airlines' preference was clear, and eventually Boeing fattened the 707's cabin by 16 inches, exceeding the width of the DC-8 by 1 inch. Then, in an inspired decision, Boeing added a higher-powered version of the 707 called the Intercontinental. Pan Am switched to this version and increased its total order to 23.

Boeing proved that flexibility won customers. The 707s built for Braniff carried a different jet engine suited to the high-altitude South American airports along the airline's routes. And the Qantas 707s were given a shorter body, to suit that airline's particular needs.

By 1962, 707s had carried 30 million people 750 million passenger miles. Boeing would eventually sell 800 aircraft, compared to a DC-8 production of 550.

Before World War II had begun, Shell Oil became the first company to employ an aircraft for corporate travel, a Fokker trimotor. But the real boom in corporate aviation came at the end of World War II, when military transports were sold to several companies who converted them into business craft. Many C-47s were rebuilt as DC-3s. Through the vision of William Lear, the Lockheed Lodestar became the Learstar.

Then, in 1963, Lear took a huge financial risk and produced a new aircraft whose sole purpose was to transport corporate executives. The Learjet exploded onto the market. Almost immediately it became the status symbol of the corporate world. Lear, a man with an elementary school education, had spurred the evolution of a lucrative industry. In his lifetime he patented more than 150 electronic and aerodynamic designs, including the eight-track tape and the dry-cell battery.

The success of the Learjet inspired many competitors. The Cessna Citation was the next big success; between 1972 and 1977, 349 Citations were sold. Though smaller and 100 miles per hour slower than the Learjet, it was less expensive, less noisy, more fuel-efficient, and allowed access to more airports. Grumman's Gulfstream I served many major corporate clients, including the Disney Corporation. Today the manufacture, maintenance, and operation of corporate aircraft has become a multibillion-dollar industry.

The Boeing 727

In 1962 Boeing introduced the 727, a jet airliner better suited to short runways. The 727 featured more efficient engines, but its selling point was a unique arrangement of wing flaps. Opening the flaps increased the wing area by 25 percent, and the plane virtually floated to a landing on difficult runways. By the 1980s, nearly two thousand 727s had been sold or ordered.

Jumbo Jets

The design of Boeing's passenger jets has always been closely aligned with the company's production of military transports and tankers. The 747, the world's first and biggest jumbo jet, was no exception.

The Boeing 727 was designed with three tail-mounted engines. This made the aircraft a popular choice with international airlines requiring extra safeguards for travel over large bodies of water. By the early 1980s, the 727 was considered the most successful commercial aircraft in history.

Adapting Airports for Jets

By the year 2000, more than one billion people will use U.S. airports each year.

Airports have changed dramatically over the past 50 years, adapting to an exponential increase in the number of passengers and to ever larger planes—from the 21-passenger DC-3s of the 1930s to the enormous, 500-passenger 747s of the 1970s. Bigger planes need longer runways for takeoff and landing. To cut down on congestion at the main terminal, airports have added corridors that telescope out to "satellites," areas that provide berths for many jets. Conveyor belts move passengers down long corridors.

Air traffic control and security systems have become far more sophisticated. Each day, the more than 200,000 takeoffs and landings at U.S. airports are monitored by air traffic controllers, who guide pilots and use radar screens to track the airplanes' movements. X-ray scanners and metal detectors are now standard security equipment, and devices are being developed to detect plastic explosives.

To study the effect of collisions with birds, testers fired chickens at the DC-8's windshield at 460 miles per hour.

The 747 In 1962 Boeing bid on a military contract to provide the transport plane that would succeed the Lockheed C-141. When the contract was lost to Lockheed, Boeing converted its design into a commercial airliner.

The proposed plane was gigantic: 19 feet, 5 inches wide, a fuselage almost as long as a football field, and a tail as tall as a six-story building. Each plane would need an entire acre of parking space.

Minimizing the massive plane's weight was a headache for designers. Engineers reshaped the wing to cut 1,000 pounds. They used heavy paper impregnated with plastic on certain non-weight-bearing external parts. Other pieces of the plane were formed of light-weight titanium.

With orders already placed, Boeing had to delay work to build a factory big enough to assemble the airplane. Pilots needed to adjust to its size, even in the simple art of taxiing around the airport. To help them, Boeing attached three-story-high stilts to a truck, and then stuck the shell of a 747 flight deck on top of the stilts. Pilots sat in the shell, directing the truck driver by radio. According to one pilot, the experience was like "sitting on the roof of my house and trying to drive the thing into the street."

Service problems cropped up as the 747 went into operation in 1970. The plane's lavatories were inadequate. Cabin attendants had difficulty serving the large number of passengers. As such problems were ironed out, jumbo jets were accepted as an economical way to travel.

The very successful 747 jet aircraft is shown under construction at a Boeing plant. The 747 was the first aircraft dubbed a "jumbo jet"; with a wingspan of almost 200 feet, a length of more than 230 feet, and seating for up to 550 passengers, it deserved the name.

The 747's size was mind-boggling to its first passengers. Its well-designed wings—with leading- and trailing-edge flaps, spoilers, and high- and low-speed ailerons—are able to lift the huge 350-ton jet.

The 747's Competitors Several new players soon entered the jumbo jet market: Lockheed, with the L-1011 TriStar; McDonnell Douglas, with the DC-10; and a European consortium, with the Airbus Industries A300.

The TriStar and DC-10 beat the Airbus into the market but could not match its long-term success. The DC-10 had been plagued by accidents and the resulting bad press. And Lockheed, failing to sell enough TriStars to cover development costs, lost $2.5 billion on the plane and left the airliner business.

The A300 managed to carve out a secure niche because its two engines, lightweight wings, and smaller size make it less expensive to operate. It succeeded in the short-haul U.S. market and has sold briskly in Europe and Asia.

Supersonic Transport

During the 1960s, Europeans were chafing at what French president Charles de Gaulle called "the American colonization of the skies." One prize remained to be won: the creation of a supersonic transport.

The Concorde In 1962, the governments of France and Great Britain decided to chase this prize together by initiating the Concorde program. Without government support, no aircraft company could join the race for an operational SST—development costs were too high. U.S. designers were forced to watch from the sidelines.

Prototypes for the Concorde established a load of 100 passengers, a cruising speed of Mach 2.2, and an operating altitude of 50,000 feet. The designers were careful not to push the plane past Mach 2.2—friction on the aircraft at that speed would have created too much heat. Materials like titanium or stainless steel could stand up to it but were too costly or too heavy.

Like many supersonic jet fighters, the Concorde was given a delta wing. The designers created a nose that could be tilted downward during landing to give the pilot a clear view of the runway. A pilot describes the Concorde's distinct look:

> After years of looking at it from every possible angle I am still not sure I can pin down its exact shape. Flying overhead at a few thousand feet it is slender, feminine, dart-like. On final approach it suggests a bird coming in to land. Just after landing, with its nose still down, it might be some prehistoric monster with curious eating habits.

Progress was dampened by negative public reaction to the noise the plane produced, including its sonic boom. The SST race was turning into a slow jog, with the Concorde as the only competitor. Hope for sales of the new plane dwindled as costs rose. No airline was willing to commit to a plane that few customers could afford to fly. Quietly, the British approached the French to discuss canceling the project. But France was committed to achieving the goal.

British Airways and Air France fly a small fleet of Concordes to over 85 destinations worldwide. Their passengers are primarily business travelers.

The Concorde Jet

The Concorde jet—the world's only supersonic transport—was developed cooperatively by British Aerospace and France's Aérospatiale. Since 1973, Concordes have transported passengers at at a cruising speed of Mach 2.2, over twice the speed of sound; the craft's speed record is 1,490 miles per hour.

The Concorde's unusual streamlined design can withstand the stresses of supersonic flight.

The small luxury cabin accommodates 100 passengers. The 204-foot fuselage stretches up to 10 inches in supersonic flight, due to heating of the airframe.

The skin is made of layers of aluminum and insulation, which protect against temperature changes.

Streamlining covers rudder control on thin tail fin.

The tail cone is part of the Concorde's aerodynamic design.

Slim, curved 84-foot delta wings carry engines and tanks for 34,000 gallons of fuel—about 1 ton of fuel per

Four turbojet engines power the Concorde. Afterburners kick in extra power for transition to supersonic flight.

The sweep of the Tupolev 144's modified delta wing lends this SST a graceful appearance on the ground—but its performance in the air left much to be desired.

Sparks of Competition A weak competitor in the SST race appeared in December 1968—the Russian Tupolev Tu-144. So closely did the plane resemble the Concorde that the British grumbled of espionage and nicknamed the plane the Concordski. But its inherent stability problems caused the Tu-144 to disintegrate in midair at the 1973 Paris Air Show, and the project was shelved.

The cold war briefly brought the United States into the SST race. Challenged by Sputnik and the Tu-144, the United States was not ready to be outclassed in another race for the skies. Boeing won the SST contract from the government but work never progressed far. Public opposition mounted, and President Richard Nixon killed the funding on May 20, 1971.

Both France and Britain had invested 14 years in the Concorde's development. It was therefore expected that the flagship airlines of these countries, BOAC and Air France, would order Concordes, which they did.

Early Concorde flights were flawless and unquestionably fast. The plane garnered a small but enthusiastic clientele who enjoyed the Machmeters on the wall and the Concorde's ability to shrink long distances into the blink of an eye. Without turning a profit, however, the Concorde did not stand a chance. The SST partners halted Concorde production in 1979, with only 16 planes built. Each carried a pricetag of $500 million. The Concorde had no peer, but it also had no market.

Vertical Flight

Early legends of flight describe vehicles that more closely resemble helicopters than airplanes—vehicles capable of vertical takeoffs and landings.

Early Experiments

Most early experimenters in flight were interested in any method possible for lifting human beings from the ground. Helicopters were a passion of Leonardo da Vinci. A full-scale drawing of a lifting airscrew (also called a helix) appears in Leonardo's papers. He mused:

> I find that if this instrument with a screw be well made—that is to say, made of linen of which the pores be stopped up with starch—and be turned swiftly, the said screw will make its spiral in the air and it will rise high.

Another early experimenter, Sir George Cayley, explored how an airscrew or rotor could provide lift. Cayley constructed several models that were probably capable of flight. The issue of power, however, remained as much a hurdle for helicopters as it had for airplanes. No propulsion system available in the nineteenth century was capable of turning the rotors.

Even the remarkable inventor Thomas Edison was brought to a standstill by the propulsion problem. In the 1880s, Edison began a series of tests to measure the lift produced by different types of rotors. He soon realized his experiments were premature: no matter what the rotor design, engines of the period could create only 160 pounds of lift.

Edison figured he could build a better engine:

> I used stock-ticker paper made into guncotton and fed the paper into the cylinder of the engine and exploded it with a spark. I got good results, but burned one of my men pretty badly and burned off some of my own hair and didn't get much further.

This drawing of an airscrew appears among Leonardo da Vinci's many sketches and notes on flight.

Igor Sikorsky pilots the tethered Vought-Sikorsky VS-300 helicopter on a test flight in 1939. Its twin-rotor design influenced helicopter builders throughout World War II.

Paul Cornu's Big Hop

Early in the twentieth century, engines powerful enough (though just barely) to lift helicopters off the ground became available. On November 13, 1907, Paul Cornu achieved the first "free flight" of a helicopter when his aircraft rose up to 5 feet off the ground for 20 seconds. Early machines such as his were typically ungainly, complicated craft. Many looked like airplane-helicopter hybrids with rotors hugely out of proportion to the rest of the craft.

Through the 1930s, experimenters throughout the world furthered helicopter theory. A fraction of the money lavished on airplanes had been provided for vertical flight, yet an industry was about to be born.

A key event was the initiation, in 1921, of the U.S. Army's first vertical flight program, headed by the helicopter pioneer and Russian engineer George de Bothezat. From the 1930s onward, military contracts and competitions would become the bread and butter of the helicopter industry in the United States. From a craft considered interesting but impractical, designers forged a powerful multipurpose machine capable of feats unmatched by any other aircraft.

The Autogyro

Another interesting pre–World War II development was the autogyro, a craft using a propeller for forward movement as well as a rotor for lift. The Spaniard Juan de la Cierva built the first successful autogyro in 1923. The man renowned as the father of the helicopter, Igor Sikorsky, claimed the autogyro was the missing link between fixed-wing craft and the helicopter.

Focke-Achgelis Models

The Germans fielded pioneers in helicopters as well as in airplane technology. One such visionary was Heinrich Focke. Through the 1930s he developed several helicopters, one of which, the Focke-Achgelis FA-61, was powered by a nose-mounted, 160-horsepower aircraft engine. Visiting Germany before World War II, Charles Lindbergh called this helicopter one of the most intriguing devices he had ever seen.

A few years later, Focke was using 1,000-horsepower engines in his machines. With that much power, the FA-223 was able to carry six people including two crew members, fly 115 miles per hour, and rise to 23,000 feet. Like the airplane, the helicopter grew up quickly, making huge leaps in capacity and taking on new roles. Another Focke-Achgelis model, the FA-330, was a surveillance gyrocopter designed to be towed by a submarine.

The Modern Helicopter

The first phase of modern helicopter development took place during the years of World War II. In the United States, Igor Sikorsky (of the Vought-Sikorsky Division of United Aircraft Corporation) was given his first contract to produce helicopters. During those heady days, when he crammed two working days into every day, Sikorsky learned to fly helicopters while testing new designs.

Thus began the R series of Sikorsky helicopters. The company delivered 600 machines, using three different designs, in a period of three years and four months. The usefulness of the machines led to enthusiastic discussions concerning new ways to deploy helicopters within the military.

Sikorsky Aircraft built 136 civil models and 481 military versions of the S-61. The military S-61, or "Sea King," was built in the 1960s. More than one hundred of the civil version (below) are still flying, filling roles such as oil-field support missions. The president of the United States flies in a variant of this S-61 medium-lift helicopter.

At an air base in Korea, an H-19 helicopter of the U.S. Air Force 3rd Air Rescue Group demonstrates the technique for rescuing a soldier behind enemy lines. The 8,000-pound, 10-passenger craft could also pick up downed air crews from the sea. Traveling at speeds up to 90 miles per hour, this helicopter had a 300-mile radius of action.

Helicopters in the Korean War

The second phase of development occurred during the Korean War. During this time, more than 50 competitors for military contracts emerged, including some of the great names in helicopter development: Frank Piasecki of Piasecki Helicopter Company, Charles Kaman of Kaman Aircraft Corporation, Stanley Hiller of Hiller Helicopter Company, and Hughes Helicopter engineers and designers.

The helicopter came into its own in Korea. Many downed pilots were pulled out of enemy territory by search-and-rescue helicopters—variously called "handy andies," "flying eggbeaters," and "gyrating angels." After medics started giving blood transfusions to wounded soldiers on helicopters, the number of soldiers who died before reaching a hospital dropped by half.

Helicopters filled many roles during the conflict: aerial staff cars, carriers of mail and blood plasma, mine detectors, reconnaissance aircraft. In one operation, 12 Sikorsky S-55s moved an entire battalion of 1,000 Marines and their equipment over 16 miles of rugged territory in 4 hours.

Engine Development

The development of gas turbine engines signaled as much a revolution in helicopter development as in airplanes. Piston engines had caused heavy vibrations and limited the helicopter's payload. With the turbine engine, the helicopter enjoyed far more power per pound of engine weight, the engine ran more efficiently while hovering, and fewer operating parts were required. Payloads rose dramatically. After 1970, the introduction of a twin-engined variant would further increase the helicopter's reliability.

Both the Bell UH-1 transport helicopter (top) and Bell AH-1 gunship (bottom) were in service in Vietnam. Losses were high—as many as 4,200 army helicopters had been downed by 1971. But U.S. combatants depended heavily on helicopters for both rescue and firepower.

The Helicopter in Vietnam

Powerful helicopter gunships, such as the Bell HueyCobra, are burned indelibly into our memories of Vietnam. In addition, much use was made of giant twin-rotor helicopters such as the Boeing CH-46 Sea Knight to carry soldiers and supplies ashore or across country. At one point in the war, U.S. helicopters were landing 770,000 tons of supplies per month.

A Multipurpose Craft

The helicopter is a unique machine, capable of flying sideways, backwards, and straight up or down. Pilots can hover motionless in the air as they lift loads and can operate from a space not much bigger than the diameter of the rotors.

These characteristics, which make the helicopter an indispensable tool to modern armies and navies, also have wide application in daily life. Hughes 500Ds carry traffic accident victims to hospitals, Sikorsky HH-3Fs check burning oil platforms for survivors, Bell JetRangers drop fire-suppressant chemicals on brush fires. Less dramatic but equally important is the use of the helicopter as a business vehicle.

Wild Ideas and Future Flight

Though many would-be aircraft inspire more laughter than flight time, there's no end in sight when it comes to creating wild flying machines. Some discarded ideas of the past—flying wings and ornithopter-style flapping wings—have become workable through the use of computer controls. Perhaps the simplest ideas are the most inspiring—they make flight available to us all.

Nicknamed the "Flying Car," Convair Model 118 ConvAirCar was designed for the post–World War II civilian market. But when the second ConvAirCar ran out of gas in flight and crash-landed, the project was abandoned.

In 1907, Alexander Graham Bell founded the Aerial Experiment Association for flight research. One disappointing result of the venture was the 950-pound honeycombed craft called Cygnet II (above).

Interesting, but . . .

The Wrights' brief soar above the sands of Kitty Hawk sparked a flurry of experimentation in flying machines. Armed with the knowledge that powered flight was possible, inventors rushed into the air. Some of the resulting designs were utterly bizarre. It is a wonder that anyone ever thought they would fly.

Alexander Graham Bell's Cygnet II was a 26-foot-wide structure made of 9,560 tetrahedral cells meant to provide lift. It looked like the front of an automobile radiator and flew just about as well. Engines of the time were not powerful enough to lift the Cygnet II, and Bell had to accept failure after a disappointing test flight in February 1909.

In 1920 an open lake in Italy was the setting for the launch of another monstrosity, the Ca 60 Transaereo. Its inventor, Gianni Caproni, knew big planes well: he had designed bombers for the Italian air force in 1914 and 1915. The Transaereo is a whimsical lapse in an otherwise brilliant aviation career.

The project started with a 77-foot houseboat, which Caproni designed to carry 100 passengers by air from Europe to New York. He figured three sets of triplane wings and eight 400-horsepower Liberty engines would lift it into the air.

On March 4, 1921, the 23-ton Transaereo did lumber 60 feet out of the waters of Lake Maggiore. Then its ballast slid into its nose, the center wings crumpled, and the plane plummeted into the lake, never to rise again.

World War II inspired two more interesting but doomed craft—one huge and one tiny. In 1942, German U-boats were blasting Allied shipping. Howard Hughes and shipping tycoon Henry Kaiser decided the answer was a fleet of enormous flying boats. Hughes talked the government into an $18 million contract for the HK-1 Hercules, but the government insisted that no "strategic materials"—meaning aluminum—be used. Hughes decided to use wood.

Hughes's engineers roamed forests looking for the best trees for specific parts of the prototype. The desire to perfect the "Spruce Goose" delayed its completion until June 1946. In November, Hughes took the 200-ton Goose up for its only flight, 60 seconds over the harbor at Long Beach, California. It reached an altitude of less than 100 feet. Hughes was so dissatisfied with the plane's performance, it was never flown again.

In 1944 the U.S. Air Force came up with the idea of a "parasite fighter"—a tiny plane to be carried under a B-36 bomber. When enemy fighters approached, the rotund XF-85 Goblin would be released to chase them away. The Goblin turned out to be too unstable.

In the 1930s, the idea of a "roadable" airplane, or flying car, was considered for the civilian market. Several different versions were proposed—some with foldable wings, others with wings that could be easily taken off and put on. Several models actually flew, but none were considered practical.

At least one aviation expert now claims that the idea of roadable airplanes may be resurrected. Imagine the traffic jams in the sky!

The crew of the Ca 60 Transaereo, nicknamed the "Capronisimo," pose proudly for a souvenir photograph on the banks of Lake Maggiore—just one week before the crash of the ungainly machine.

The Hughes-Kaiser HK-1 Hercules, or the "Spruce Goose," (left) was considered obsolete before it was finished. When the U.S. government grant of $18 million ran out, Hughes invested $7 million of his own into this pet project.

Wondrous and Workable

Today airplane designers still come up with wild new ideas—and many of them work brilliantly.

Homebuilts and Ultralights

Recreational flying is not limited to lightplanes produced by companies like Piper and Cessna. Today's backyard airplane builders have made the homebuilt movement a source of astonishing new ideas in aviation.

Homebuilts, small airplanes built by the pilot, include popular kits such as Burt Rutan's VariEze and Long-EZ. Smaller versions of classic planes, such as World War I biplanes, also fall into the homebuilt category. Ultralights, originally hang gliders powered by small engines, also enjoy an enthusiastic following. Each new generation of homebuilts includes faster, more complex machines.

The lively interest in this realm of experimental aviation is shown by the tens of thousands of people who crowd the Experimental Aircraft Association's annual air show at Oshkosh, Wisconsin, and the National Championship Air Races in Reno, Nevada.

Dr. Paul MacCready heads the engineering firm of AeroVironment Inc. in Monrovia, California. The company specializes in ultra-efficient air and land vehicles.

Composites Composite materials have revolutionized the construction of homebuilts and are becoming an important component of all types of aircraft. Composites combine substances such as metal, ceramics, and plastics to create materials with greater strength or flexibility than each possesses alone. Composites such as Kevlar, carbon and glass with foam, honeycomb plastics, and graphite fibers bound with epoxy can form stronger, smoother surfaces that reduce drag and make previously unworkable ideas possible—for example, long, thin wings or swept-forward wings.

The Flying Machines of Burt Rutan and Paul MacCready

Designers such as Burt Rutan and Paul MacCready have influenced aviation with their imaginative designs. They have produced innovative, captivating aircraft. Both are known for unusual record-setting airplanes.

Rutan's *Voyager* The flight of the airplane *Voyager* grew out of ideas scribbled on a dinner napkin in 1981. Pilot Jeana Yeager and the brothers Burt and Dick Rutan were meeting to discuss new ventures. Burt proposed that he build the first plane capable of flying around the world without refueling. Dick, an ex–fighter pilot, and Jeana would crew the ambitious venture.

Ultralights

An ultralight weighs less than 254 pounds, carries no more than 5 gallons of fuel, has only one seat, and attains a maximum speed of 63 miles per hour. Ultralights appeared during the hang-gliding movement of the 1970s. Mid-westerners had no cliffs to launch from, so they strapped motors to their hang gliders and launched them by foot, running until they lifted into the air.

Ultralights are inexpensive and fly slowly and close to the ground, letting the pilot savor flight.

Since mid-1993, ultralight aircraft have come under a new category of certification by the FAA. The first Primary Category type certificate went to the ultralight sportplane Quicksilver GT 500 (left). People in the light aviation industry expect the new, simpler rules of certification to rejuvenate the business.

As Dick and Jeana began raising funds, Burt focused on the airplane's design. Constructed using an ultralight graphite composite, *Voyager* featured a small forward set of wings called *canards*, which act as stabilizers and provide some lift, and vertical winglets on the tips of each wing, which improve the efficiency of the wing.

Burt then faced the same problem confronted by long-distance aviators since Lindbergh. Fuel was a vital element in the flight's success; the penalty for running out was obvious. But how to carry enough fuel to fly around the world, along with a safety reserve?

To create enough space for the 1,489 gallons of gasoline, fuel tanks were built into the long, knifelike wings. More space for fuel was gained by placing twin tail booms on either side of the central fuselage. In front of the wing, canards joined the booms to the slender fuselage.

The *Voyager* flight broke aviation records by completing a nine-day, nonstop, unrefueled flight around the world. The aircraft cruised below 11,000 feet for most of the flight, but, to avoid turbulence over the African continent, Voyager reached altitudes over 20,000 feet.

But the heavy fuel endangered the plane during takeoff and on the first leg of the journey. Gasoline would make up three-fourths of the *Voyager*'s total weight. Under that heavy load the wings would flex as much as 10 feet on takeoff. Burt chose to move ahead with only a slim 10-percent safety margin between successful flight and catastrophic wing failure.

Dick and Jeana had trouble finding sponsors willing to put their name on a craft that might crash on the runway. Though corporate sponsors eventually signed on, devoted volunteers were responsible for building, testing, and repairing *Voyager*. Four years later this same group planned the route and monitored the flight's progress.

Flight tests did not lessen concerns about the plane's performance. Even under a partial fuel load, *Voyager* was difficult to fly. No attempt was made to test the plane with a full load.

On December 14, 1986, Dick and Jeana climbed aboard to begin the flight. For a few harrowing moments the flight seemed in jeopardy, as *Voyager*'s fuel-heavy wings dragged across the tarmac. Then Dick eased the plane into the air, beginning nine days of boredom, exhilaration, anxiety, difficult weather, and sleep deprivation in the airplane's cramped cabin.

Nine days later at Edwards Air Force Base, 50,000 people greeted *Voyager*'s triumphant return. Jeana Yeager and Dick Rutan had flown around the world without refueling.

MacCready's Flying Machines One mark of a Paul MacCready design is the unusual fuel source. The pilot's leg muscles, for example, provide the push for the *Gossamer Condor* and the *Gossamer Albatross*. The *Condor* won the Kremer Prize for human-powered flight, and the *Albatross*, piloted and pedaled by Bryan Allen, won the Kremer for crossing the English Channel.

The *Gossamer Penguin,* a three-quarter-scale version of the *Albatross,* is powered by electricity from solar cells. The pilot's compartment in the *Penguin* hangs beneath the wing. Above it is a bank of solar cells positioned on two pylons. MacCready's *Solar Challenger* also crossed the English Channel.

Another MacCready creation is a mechanical model of a giant Cretaceous pterosaur, the *Quetzalcoatlus northropi.* The wing-flapping, tailless model shows the effect of computers on aviation—though aerodynamically unstable, *Quetzalcoatlus* can fly because of its sophisticated computer controls.

Paul MacCready's prize-winning *Gossamer Condor* was assembled from 55 pounds of plastic, wire, and aluminum tubing. Housed below a 96-foot wing was a modified bicycle frame attached to a 13-foot propeller.

Today's State-of-the-Art Aircraft

The ongoing revolution in computers, materials, and aerodynamics has forged a new standard for every type of airplane. Each plane described here bears the mark of that revolution.

Yet calling these aircraft "state-of-the-art" is somewhat misleading. Military and commercial aircraft suffer from long lead times in development. Even the most current design of each aircraft was frozen in the 1970s, when manufacture and flight testing began. Also, the astronomical costs of new airplane design can be a deterrent to innovation. It's safer to stick to the tried-and-true.

Fighters

The flashy supersonic fighters of today carry amazing price tags. A single fighter can cost $30 million and weigh 25 tons. Compare that to a World War I fighter, which could cost about $50,000 and weighed about half a ton.

Electronics make up at least one-quarter of the cost of a modern fighter. Pilots still rely heavily on their eyes during dogfights, but radar provides the winning edge during a "merge" with the enemy. The electronics for radar alone can cost well over $1 million. Other systems cover electronic countermeasures, command and control, navigation, and landing.

The X-31

The X-31 maneuvers by stalling. In the past, pilots have always feared a stall because the wings no longer produce lift. Stalling, said one developer, is like disconnecting a car's steering wheel at 65 miles per hour.

Through an intentional stall, however, the X-31 can slow down, and thus increase its agility. After stalling, the pilot uses a technique called thrust-vectoring to kick the tail of the X-31 around and roll into an abrupt about-face. This maneuver lets the pilot make turns in half the space required by other aircraft—enough of a difference to rewrite the rules of fighter combat.

The Skunk Works

The billion-dollar SR-71 (above) has made highly classified surveillance missions. On May 1, 1960, a U-2 spy plane (below) was shot down over the Soviet Union.

In 1943 a highly creative designer named Clarence "Kelly" Johnson founded the Skunk Works as a research arm of Lockheed. The nickname "Skunk Works" was inspired by a character in the comic strip *Li'l Abner,* who was avoided because he was "inside man at the Skunk Works." The success of the group has turned "skunkiness" into an aviation techno-religion.. Projects cloaked in secrecy are called "black and skunky."

During the cold war the CIA gave Johnson a blank check to develop two spy planes. In August 1955—nine months after the contract was signed—the first U-2 spy plane was in the air.

Kelly Johnson next started working on the U-2's successor, the SR-71 Blackbird. First flown in 1962, the SR-71 was capable of reaching 90,000 feet and a speed of Mach 3.3. The SR-71 is nicknamed "Habu," the name of a black, poisonous snake of Okinawa. It holds several speed records, including a 64-minute flight between Los Angeles and Washington, D.C.

The Skunk Works' most recent success is the stealth fighter, the F-117A Nighthawk, praised by pilots in the Persian Gulf War. The next challenge is to build a hypersonic airplane capable of reaching Mach 26.

Like the X-31, new models of the F-16 Falcon are being equipped with multi-axis thrust vectoring (MATV). Adding the thrust-vectoring system has not required structural modifications to the airframe.

Half the price of a plane goes toward its airframe. The skin of modern fighters is formed from an expensive mixture of rare metals (such as titanium), alloys and superalloys, and custom composites. The engine is the single most expensive item in a fighter, costing as much as $3 million.

Due to these numbers, and to each fighter's increased firepower, air forces now own hundreds of fighters, not thousands. Also, many countries resort to buying rather than building fighters. Not all are willing to follow the United States, which has at times spent 26 percent of its entire defense budget on electronic hardware.

The F-15 Eagle is regarded as the U.S. Air Force's leading long-range fighter. F-15 Eagles can carry 24,500 pounds of ordnance (military supplies including weapons) and were deployed in the 1990–1991 Persian Gulf War.

The General Dynamics F-16 After intense international competition among manufacturers, many NATO countries adopted the General Dynamics F-16. Called the "arms contract of the century," sales of the fighter in the 1990s are predicted to reach $15 billion. The F-16 was an experiment, a low-cost, lightweight fighter intended to supplement the heavier, more expensive McDonnell Douglas F-15. It is capable of attacking both air and ground targets.

The F-16 (now built by Lockheed, which purchased General Dynamics) is a showcase for many new features—composites in the airframe, a special canopy for high visibility, and a cockpit designed to improve the pilot's tolerance for high G forces. Its fly-by-wire flight-control system employs direct electrical signals, rather than mechanical connections, to move controls. It is known as the Electric Airplane.

The McDonnell Douglas F/A-18 Hornet is a major export earner for the United States. In the Persian Gulf War, carrier-based Hornets were used in air combat and mid-range ground attack. Their weaponry included Harpoon and Maverick missiles.

The AV-8B Harrier II is used extensively by the U.S. Marine Corps for training exercises on board assault carriers. Capable of firing missiles, cluster bombs, or smart bombs, the jet was successfully deployed as a ground-support jet in the Persian Gulf War. Fitted with thrust-vector engines, the Harrier can decelerate more rapidly than any other aircraft.

VTOL Aircraft

Newly developed engines have improved the performance of another type of military craft—Vertical Takeoff and Landing (VTOL). These airplanes take off and land vertically like a helicopter, but fly like conventional aircraft in the air. This vertical movement requires an engine that provides enormous thrust.

The AV-8B Harrier Though the VTOL concept dates back to the 1960s, the first operational VTOL craft is the British Aerospace Harrier, which first flew in 1981. The new technology that makes the old idea work is called *thrust-vectoring*. Harrier pilots can rotate the four nozzles of the jet engine to provide thrust in different directions. An improved model is the AV-8B produced by British Aerospace-McDonnell Douglas.

Engine design is not the only new technology in the AV-8B. A graphite-and-epoxy composite makes up over a quarter of its airframe. Improved wing design allows greater lift, more efficient cruising, and a greater fuel capacity. And the fighter, which is flown by both the U.S. Marine Corps and the British military, boasts a cockpit filled with modern electronics.

The Latest Features

New Wings

At first glance, some modern wing designs look quite odd. Why would anyone, for example, think that a small vertical winglet on a wing's tip would improve its performance? The answer lies in a completely new method of aircraft design.

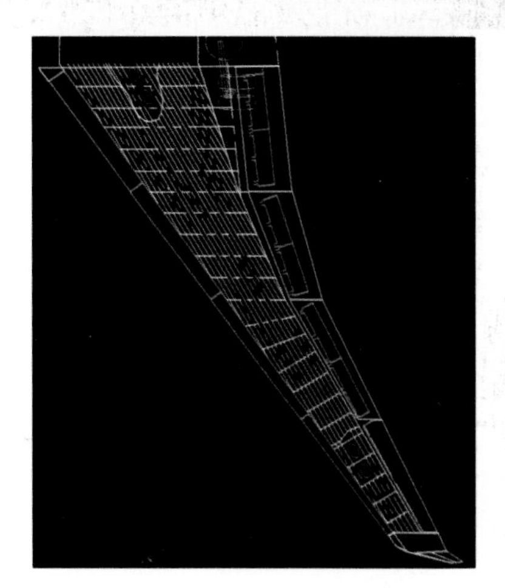

This aircraft wing was designed by computer—only one of many technological advances revolutionizing aviation. Robotics and artificial intelligence are other innovations being used by the aircraft industry to cut costs and streamline design.

Computers are now used to test new concepts. Engineers perform on-screen tests in minutes that previously would have taken months of expensive wind tunnel work or flight testing. They can experiment with hundreds of changes to a wing, making corrections by keyboard or light pen. Computer-aided design (CAD) has contributed speed and efficiency to an engineer's work. Some of the results are described here.

The Flying Wing John Northrop first conceived the idea of a plane built entirely from wing surfaces in the 1940s. Wings provide lift, but the rest of the airframe contributes to drag. Therefore, he reasoned, a plane's aerodynamics would be improved if the weight were placed where the lift is—along the wingspan.

Northrop and the U.S. Air Force built several flying wings, including the piston-powered XB-35 and the jet-powered YB-49. Though the planes flew, the design proved difficult to control in flight without conventional tail assemblies.

That stability problem has been solved by improved control systems. Now the flying wing concept is reappearing in military aircraft and is being considered for large cargo transports. Since it has no large vertical surfaces to reflect radar signals, its shape is a key factor in stealth aircraft, offering "low observables."

The Northrop N1M Flying Wing is a high performance wing featuring a total of 28 composite control surfaces, 14 on each side.

Stealth Bomber and Fighter

Stealth aircraft are able to avoid radar detection. Their unusual shapes don't resemble typical airplanes at all. Dark coatings, peculiar angles, and hidden engines and armament give stealth craft a mysterious, almost surreal appearance.

Unveiled in 1988, the B-2 stealth bomber (above) employs a flying-wing design. First conceived by John Northrop in the 1940s, the wing has gained greater flight stability with computer controls.

Nicknamed the "Wobbly Goblin," the F-117A Nighthawk stealth fighter (below) was secretly tested throughout the 1980s and first saw service in 1989. It excels at nighttime precision attacks on high priority targets.

The flying wing design has minimal vertical surfaces, enabling it to escape detection by radar.

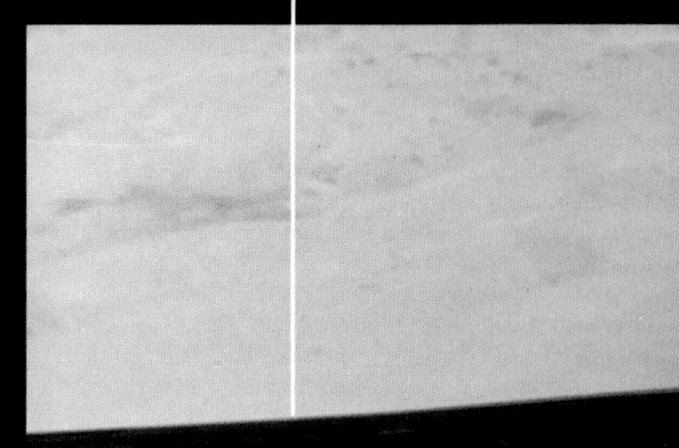

Angled panels deflect radar energy away from the fighter.

The F-117's two engines are embedded deep within the wingroots, reducing detection of engine heat.

Air-to-surface missiles are stowed within the fuselage.

Engine intake and exhaust ports are designed to reduce infrared and radar image.

Cockpit holds crew of two.

Four turbofan engines push the bomber to a maximum speed of 700 miles per hour.

Two weapons bays can carry nuclear or conventional bombs.

The B-2 bomber is coated with dark, flat material that absorbs 99 percent of radar energy rather than reflecting it, thus helping to prevent detection.

A video system can be used for takeoff on blacked-out airfields; an infrared system and laser sights are used for targeting.

Fly-by-wire controls make this unstable flyer maneuverable and even aerobatic.

The Grumman X-29 is considered the best-performing aircraft to use a swept-forward wing. The X-29 has demonstrated a test speed of Mach 1.46.

Swept-Forward Wings Swept-back wings took the world of aviation by storm after World War II. Allied engineers were introduced to German data proving that sweeping wings backward decreased drag on a plane traveling at high speeds. Planes with swept-back wings appeared in the air a few years later.

German engineers also experimented with swept-forward wings during the war but with no success. The materials used at the time were not strong enough, and the wings twisted.

Now, composites are manufactured specifically to handle wing stresses. Swept-forward wings provide the same drag reduction as swept-back wings, but they allow better roll control at low speeds.

Mission-Adaptive Wings Most military aircraft perform a wide range of missions, and one wing shape is not suited to them all. Using an on-board computer, the pilot can change the shape of a mission-adaptive wing during flight, exactly as a bird does. The surface contours of a wing can be varied on demand.

Canards These small forward wings have become a familiar feature in airplanes, popularized through their use in *Voyager* and other Burt Rutan designs. Canards also grace military research aircraft such as the X-29. They have appeared on experimental craft throughout the history of aviation.

Canards provide lift, maneuverability, and some stability. They take the place of horizontal stabilizers (elevators), which

Computer-controlled canards grace the Beechcraft Starship 2000A—a small light aircraft designed for executive transport.

often create negative lift, forcing designers to use larger wings to compensate, which adds weight to the plane.

In recent experiments by Lockheed and the National Aeronautics and Space Administration (NASA), a vertical canard is being used to enable a plane to fly sideways.

Winglets and Wingtip-Vortex Turbines A wing produces lift because of a difference in pressure on its upper and lower surfaces. A swirl of air churns at a wing's tip, where air flows from the lower to the upper surface. The more lift a wing produces, the stronger the *wingtip vortex*. Wingtip vortices increase drag, decrease lift, and leave a wake like that of a speedboat to buffet other aircraft.

Richard Whitcomb, of NASA, devised a solution to these problems: winglets, a remarkably simple idea that is becoming a common design feature of aircraft. These vertical wing extensions, set at precisely determined angles on the main wing, improve wing efficiency and reduce fuel consumption by decreasing wingtip vortices.

Turbines take the winglet concept one step further. Instead of winglets, engineers add a turbine to the tip of each wing. The turbine has two functions: it reduces the strength and drag of the vortex and extracts energy from the swirling air. Once the turbine is connected to a generator, the energy can be converted into electricity to be used by the aircraft.

Oblique Wings and X-Wings Though swept-back wings work well for high-speed flight, they do not provide enough lift to allow a slow approach and landing speed. NASA's R. T. Jones found that a wing that pivots around its center point can work for both speeds of flight. The oblique wing is a slender, continuous piece of composite material. Held at right angles to the fuselage during takeoffs and landings, the wing is pivoted up to 60 degrees during supersonic flight. When placed at an oblique angle, it creates the same effect as a swept-back wing.

An X-wing is shaped like two long wings crossed at right angles. At low speeds, an X-wing operates like a helicopter rotor to help propel the plane. When greater speed is needed, the wing is locked into place, and the airplane flies like any fixed-wing craft.

Though X-wings and oblique wings are experimental, some airplanes, such as the F-14 and the B-1, incorporate a wing, known as a "swing-wing," that can be swept to different positions.

The makers of the Cessna CitationJet, marketed toward business customers at an introductory price of just under $3 million, claim that the jet is easier to fly than any twin-propeller plane, while offering fuel efficiency and requiring only a short runway.

Laminar Flow Control The thin sheet of air next to the surface of a wing is called the *boundary layer*. At low speeds, the flow of air near the wing is termed *laminar* because the air moves smoothly in well-ordered layers. At higher speeds, however, the flow becomes turbulent, with air molecules mixing between layers. This increases surface friction and drag.

Designers have developed ways to keep air flow smooth in the boundary layer. Sucking boundary air through holes or slots in the wing surface is one option; using a porous material for the wing surface is another. Pumps suck the air through the wing surface and vent it through ducts.

Yet another possibility is natural laminar flow (NLF). By carefully designing the shape of the wing with the aid of computers and making its surface extremely smooth with the use of composites, engineers can delay the transition from laminar to turbulent flow. The Cessna CitationJet is equipped with wings that provide natural laminar flow.

Reducing drag through laminar flow control is well worth the effort. Fuel consumption in commercial airlines could be reduced by as much as one-fourth by using this technique.

Propulsion

Throughout the history of aviation, improved engines have been the key technology that allowed new records to be set.

Hypersonic Flight Hypersonic flight—which includes any speed over Mach 5—is a challenging goal. The aircraft must fly at very high speeds and operate at very high temperatures. Yet it must be able to fly slowly enough to take off and land on conventional airport runways. To operate at such extremes—from runway speeds to Mach 12—requires a "combined-cycle engine"—one that could operate in different ways at different speeds.

Turbojets, Ramjets, and Scramjets At low speeds, one type of combined-cycle engine would act like a conventional turbojet. Turbojets slow down incoming airflow in order to convert the air's velocity to pressure and thereby provide thrust.

Air heats up as it is slowed down. As turbojets reach Mach 3 to 3.5, the amount of deceleration necessary for combustion makes the air hot enough to melt the engine.

Because it has no moving parts, a ramjet is able to partially overcome this air temperature problem. A ramjet has no compressor; the ramming effect of high-speed air at supersonic speeds provides compression. This combined-cycle engine would operate as a ramjet as it accelerated beyond Mach 3. The top speed suitable for a ramjet is from Mach 5 to 6.

Truly hypersonic flight is only possible with a scramjet. In a scramjet, air is mixed with fuel and ignited while it is still traveling at supersonic speeds. Because the air does not have to be decelerated for combustion to take place, it remains much cooler. Hydrogen is the only fuel capable of igniting fast enough to work in a scramjet.

Improved wind tunnels and supercomputers have allowed more intensive study of the capabilities of the scramjet. Current tests suggest that the scramjet could operate at speeds up to Mach 14—well into the hypersonic range and on the boundary between aviation and spaceflight.

Propfans and Fanjets Saving fuel has motivated a renewed interest in the use of propellers as a source of power. Propeller blades much shorter than those of earlier airplanes, made of composites or alloys in a swept-back shape, combine high speed and fuel efficiency.

Development of the propfan (also called the unducted fan), with short, swept-back blades, was begun in response to the high fuel prices of the OPEC crisis in the 1970s. It has been tested successfully, but practical applications of the propfan have not yet appeared.

Fanjet—multiblade, ducted propellers—offer efficient propulsion as well as decreased noise; they are widely used in both military and commercial craft. Fanjets demonstrate the fuel efficiency that propellers provide—they use 25 to 30 percent less fuel than advanced turbofan jet engines, yet deliver comparable power.

In concert with the aircraft industry, the National Aeronautics and Space Administration (NASA) is working to improve jet efficiency. Propfans (below) were proposed in response to the high fuel prices during the OPEC crisis.

Flight Control and Avionics

Military pilots in a dogfight, commercial pilots in a wind shear (a dangerous downdraft), private pilots faced with conditions beyond their training: all need help organizing and interpreting data quickly. Avionics—electronic systems and equipment developed for flying—provide that help through a complex array of computer screens, graphics, and controls.

Head-up Display (HUD) A transparent HUD screen, superimposed on the aircraft's windscreen, places all key information at eye level. A pilot need never look down or to the side to find information.

In the center of the screen is a level indicator, showing the position of the airplane relative to the earth's horizon. Other information, such as the craft's speed and altitude, is located near the level indicator. In fighter planes the HUD replaces earlier aiming mechanisms for weapons. A moving circle on the screen tracks heat sources for missile attack.

HUDs are now found in the cockpits of airliners and fighters. Scaled-down models may soon appear in business and pleasure craft.

The Head-up Display (HUD) is now standard equipment in fighter aircraft. At eye level, HUD supplies all information necessary for flying the plane, freeing the crew from checking various instruments in the cockpit.

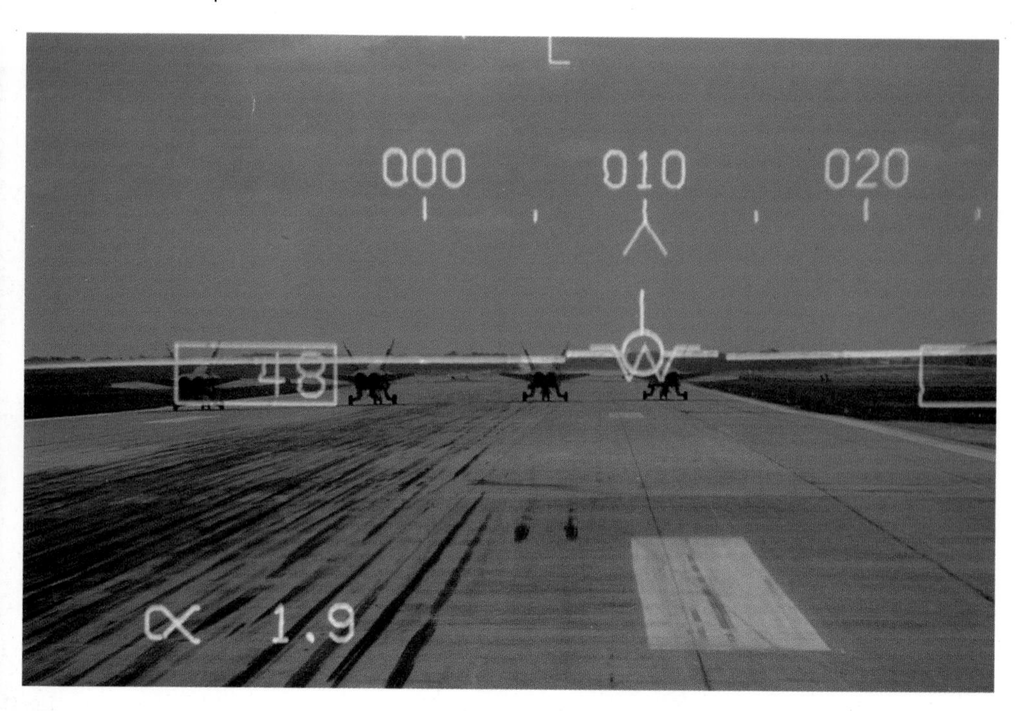

At the Threshold of Spaceflight

The North American X-15, tested by pilot Scott Crossfield, achieved altitude records that far exceeded its designers' expectations.

The first airplane to approach the threshold of space was the X-15, built by North American Aviation in the late 1950s. Made from Inconel-X, the heat resistant alloy used for jet engine turbine blades, the X-15s were intended to explore hypersonic flight.

Unintentionally, the rocket-propelled X-15 laid the groundwork for piloted spaceflight by flying well beyond its limits. By 1963, its skin was able to withstand a scorching temperature of 1,300 degrees Fahrenheit. The planes were soaring to 354,200 feet—100,000 feet above their ceiling—a still-standing world altitude record for winged aircraft. Five of the military X-15 pilots were later awarded astronaut wings.

During tests in 1967, an X-15 pilot broke through to Mach 6.7. Temperatures of 3,000 degrees burned off the scramjet and created a hole in the fuselage. Badly damaged, the X-15 landed, having set the world speed record for a winged flying machine.

Another 1950s technology still considered state-of-the-art is the U-2 spy plane now called the ER-2. Its pilots suit up like astronauts because the plane operates above 95 percent of the earth's atmosphere. The ER-2 now carries scientific instruments and gathers data on snow cover in the Rockies, damage to coral in the Florida Keys, the expanding ozone hole and other information.

Virtual Cockpits With virtual cockpit technology, the HUD is moved to the pilot's helmet. There, the information is organized both in time and in three-dimensional virtual space. Miniature cathode-ray tubes produce the display, which is then projected through the visor optics into the pilot's field of view.

When the pilot's head is turned, the informational display moves with it. Fighter pilots would be able to detect, track, and destroy targets by voice command.

Magic Windows Another possible extension of this technology is the "magic window." Forward-looking infrared cameras could be used to produce images of the scene outside. These cameras, unlike the human eye, are not blinded by rain, fog, snow, or darkness. They could provide a picture of the world outside until the weather cleared enough to allow the pilot good visibility.

HOTAS HOTAS (hands-on throttle and stick) is a control stick and throttle with buttons that control virtually every flying system a pilot uses—steering and maneuvering, weapons systems, propulsion, and so on. It reduces workload and enables a pilot to react quickly without having to look down to operate switches.

Fly-by-Light Systems In these systems, optical fibers pass signals to the plane's control systems. Using light from a laser, these signals pulse along the fibers at the speed of light. Just as fly-by-wire systems are less bulky than cranks and pulleys, optical fibers are lighter and less bulky than wiring. A tremendous amount of information can be carried through a fiber as thin as a human hair.

Supermaneuverability There is no such thing as a fighter aircraft that is too maneuverable. Supermaneuverability refers to two new capabilities: movement at high angles of attack and sideways movement.

The vectored-thrust engine of the VTOL AV-8B Harrier gives this fighter some supermaneuverability. The Harrier can, for example, slow down so quickly that it appears to have stopped in midair. A pursuing aircraft would fly straight past, suddenly becoming the hunted rather than the hunter.

Other research on supermaneuverability has been conducted with a modified F-16. The airplane's flying characteristics changed dramatically with the addition of 8-square-foot canards. Suddenly, the F-16 could move sideways without banking, rolling, or changing the direction of its nose.

Coming Soon—or Already Here

Aircraft now emerging from the drawing boards show the speed at which change is occurring in computers, materials, and aerodynamics. As always, the designs of these craft are a mixture of old ideas percolating through the market and new concepts beyond anyone's imagining even a decade ago.

The New Airship

Airships perform one task better than anything else on land or sea, in the sky or in space: long-duration, low-altitude surveillance. Dressed up with new engines and with gondolas made of composites, airships can carry the largest radar systems with ease and can stay on station for weeks at a time, detecting low-flying missiles or drug-running aircraft.

Traffic along the south Florida coast is already monitored by 250,000-cubic-foot-capacity airships nicknamed Fat Alberts. They float at approximately 11,000 feet, tethered by a 25,000-foot cord. A launch-control vehicle reels the tether in and out. A Fat Albert's radar can pick out targets up to 150 miles away.

This "Fat Albert" is used by U.S. Customs, Miami, Florida, to monitor air traffic. Airships, which are especially valuable for low-altitude surveillance, are also used by the U.S. Navy and the Coast Guard.

Cockpit of McDonnell Douglas F/A-18 Hornet

The cockpit of the F/A-18 Hornet is designed to give the pilot all the important information needed to function efficiently in the high-speed, high-g environment of air combat. New technologies are used to simplify the information from the aircraft's systems. Still, the cockpit's standby instruments can be easily recognized as descendants of Lindbergh's instruments in the *Spirit of St. Louis*.

Head-up display (HUD)
A/A
A/G
BRT
DIM
EM
CON
VOL
OFF
COMM
2
ILS
D/L
BCN
ON
OFF
BRT
Right digital display indicator
Attitude indicator
AUTO
DAY
Electronic warfare (EW) display
Standby airspeed indicator
Standby altimeter
Multipurpose color display (radar, navigation)

Military Aircraft

Advances in electronics promise to extend the life span of fighters now coming into the market. Sweden's new Gripen fighter, for example, is a multitask aircraft used for interception, ground attack, and reconnaissance. By switching computer programs and weapons, the Gripen can be adapted to new roles.

The Lockheed F-22 Another much-heralded aircraft is the advanced tactical fighter (ATF), soon to enter fighter squadrons as the Lockheed F-22. Called the successor to the F-15 and F-16, the F-22 is expected to be the backbone of the U.S. fighter force for the next 30 years.

Because of security requirements, much of the aircraft's design remains veiled. Still, some of the advances that give the airplane its unprecedented capability can be outlined here. A greater portion of the F-22's airframe and wings is formed from advanced composites than in any previous fighter. These materials include the thermoplastics and radar-absorbing materials needed to make the F-22 "stealthy."

The F-22 can be commanded by the pilot's voice. A digital flight-control system (DFCS) executes these commands by moving control surfaces as often as 40 times per second. The DFCS computer also can be programmed to act as a back-seat driver, ensuring that the pilot does not rip the plane's wings off pulling out of a dive or fracture the fuselage by turning too sharply. Artificial intelligence systems help select weapons and analyze flight data.

In the "all-glass" cockpit, dials and gauges are nowhere to be seen, replaced by color computer displays. Among these may be a "God's eye view," showing the entire combat area as it would be viewed from above the aircraft.

The engine is perhaps the greatest innovation in the F-22. It provides up to twice as much thrust per pound at supersonic speeds as the engines in the F-15. Because the plane can cruise at supersonic speeds without using an afterburner, its infrared signature is greatly reduced.

The next fighters down the line will carry the advances of the F-22 even further. The composites in the airframes will cut fighter weight dramatically and keep the aircraft cool even at supersonic speeds. Sensors in the planes' "smart skin" will relay data to cockpit computers. Cockpits will boast a three-dimensional spherical data display that will make HUD look antique.

Other possible advances include self-repairing control systems, or systems that can sense when the pilot is unconscious. Engines in these fighters will contain 1,000 to 2,000 parts, instead of the 15,000 to 20,000 parts common today. Manufactured with advanced materials, these engines will be light enough to routinely achieve 20-to-1 thrust-to-weight ratios.

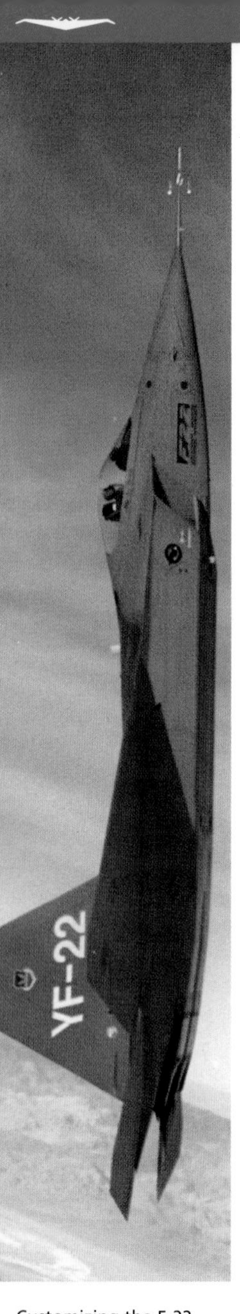

Customizing the F-22 to meet U.S. Air Force requirements will cost $20–25 million. The first flight of the eagerly awaited F-22 is scheduled for early 1997.

Air Freighters

Though only a few routes in the world can keep jumbo jets at full capacity, a huge market invites the development of new air freighters. Designs proposed for these craft appear unusual, to say the least. One concept, called a multibody carrier, calls for joining two fuselages from currently produced airliners. The tails of the two fuselages would be joined at the tips for stability.

Another visually intriguing idea is the flatbed freighter. This concept merges an airplane with a flatbed tractor-trailer. Behind the short fuselage, containing the crew compartment, would lie a completely flat cargo area. Outsized cargo such as construction equipment could be lashed down on this area and carried open to the elements. Containerized cargo would be another option. These containers could be designed to fit together into a streamlined fuselage, thus reducing drag.

This Canadair CL-227 Sentinel is a remotely piloted, rotary-winged system powered by a small Williams International gas turbine. The Sentinel can remain airborne for up to four hours.

Remotely Piloted Vehicles (RPVs)

Using unpiloted RPVs for reconnaissance and surveillance saves pilots' lives. Furthermore, the tiny and virtually noiseless RPVs can sneak up on enemies in a fashion unmatched by piloted, high-speed spy planes.

The U.S. Army's Aquila, an RPV weighing 260 pounds and made of the composite Kevlar, can fly up to 114 miles per hour at altitudes up to 12,000 feet. It is launched from a hydraulically operated catapult, mounted on a 5-ton truck, and recovered by flying the machine into a vertical net. Efforts are currently under way to develop an RPV that operates like a helicopter.

The Hunter (left), a tactical unpiloted aerial vehicle, is a joint effort of the U.S. company TRW Avionics & Surveillance Group, and Israel Aircraft Industries.

The Bell-Boeing V-22 Osprey is a tilt-rotor helicopter designed for multipurpose military use. It was first flown in March 1989.

Tilt-Rotor V/STOL

All branches of the U.S. armed forces have been eagerly awaiting delivery of the V-22 Osprey. This craft takes off like a helicopter but flies like a turboprop, with speeds exceeding 300 miles per hour. Airframe composites have significantly reduced the weight of this V/STOL (Vertical/Short Takeoff and Landing), which is suitable for combat assault, search and rescue, and medical evacuation missions.

Commercial Aircraft

The Boeing 777 The newly unveiled Boeing 777, the world's largest twin-engine jet, completed its first flight in June 1994, and more than 200 orders already have been placed for it.

The first three 777s, built for flight testing, can measure performance data on 40,000 flight characteristics.

Boeing's 777 is one of the most thoroughly tested commercial airplanes, with more than three times the hours of testing logged for the 767. It will carry up to 440 passengers with a range of up to 4,560 miles.

Design of the plane was managed by a program named CATIA (computer-aided three-dimensional interactive application), licensed by IBM from Dassault in France. Engineers sat at two thousand workstations connected to a central complex of eight supercomputers, which could process three trillion bytes of data. Graphic displays showed the 777's systems in three dimensions, color coded for easy recognition. Mistakes were corrected before, not after, the plane was built.

The 777 is Boeing's answer to its main competition, the Airbus 330 and 340, which have doubled their market share since 1990. Boeing spent $4 billion on its newest jumbo jet, and intense commercial interest in the aircraft has caused friction in the industry. In mid-1994, Boeing took legal action against Airbus for patent infringement of the 777's wing design. Tough competition for this multi-billion-dollar market led one analyst to comment: "This battle is going to get a lot bloodier."

This view of a U.S. space shuttle, carried on top of a Boeing 747, portrays the progress from flight to spaceflight achieved in the twentieth century.

Aeromarine

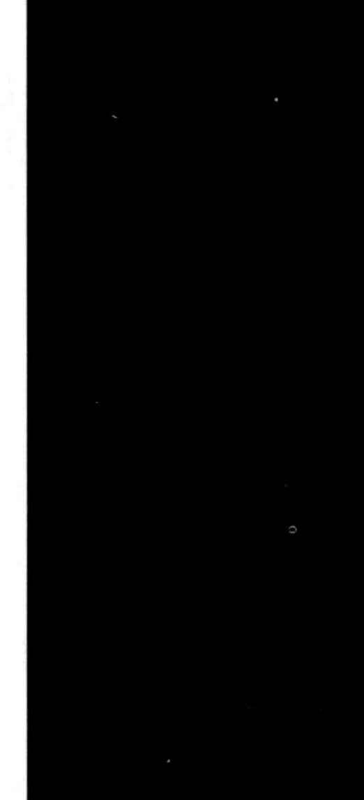

A

ace A combat pilot who has brought down at least five enemy planes.

Ader's *Avion III*

Ader, Clément (1841–1926) A French engineer and inventor. Ader's inventions include a microphone and a public-address device, but he is best known for his steam-powered, bat-winged monoplane, the *Eole,* which reportedly hopped into the air in 1890. Its flight demonstrated that a piloted, self-powered aircraft could successfully take off from level ground.

Aerial Experiment Association A group organized to research heavier-than-air flight. The group was organized and financed by Alexander Graham Bell and used his estate in Nova Scotia as a test site. Glenn Curtiss was one of the members of the group. They designed, built, and tested several airplanes in 1908 and 1909.

Aeroflot The state airline of the Soviet Union, founded in 1928 as Dobroflot and reorganized as Aeroflot in 1932. Before the dissolution of the Soviet Union, Aeroflot was the world's largest airline, making up about 15 percent of civil air traffic worldwide. Aeroflot continues now as a Russian-based airline.

Aeromarine West Indies Airways The first American airline to offer scheduled international passenger service, from several U.S. cities to Havana, Cuba, and the Bahamas. The service, which began in 1919, was nicknamed the "Highball Express," because many of its passengers sought relief in the tropics from Prohibition in the United States. The airline folded in 1923.

aeronautics The art or science of heavier-than-air flight.

Aeronca Begun as the Aeronautical Corporation of America in 1928 (and renamed in 1941), this company built and sold the first commercially successful lightplanes, becoming one of the pioneer manufacturers of light aircraft. In the 1950s Aeronca stopped making light planes and began subcontract work for larger manufacturers.

afterburner A burner attached within the tail pipe of a turbojet en-

Aircraft Carrier

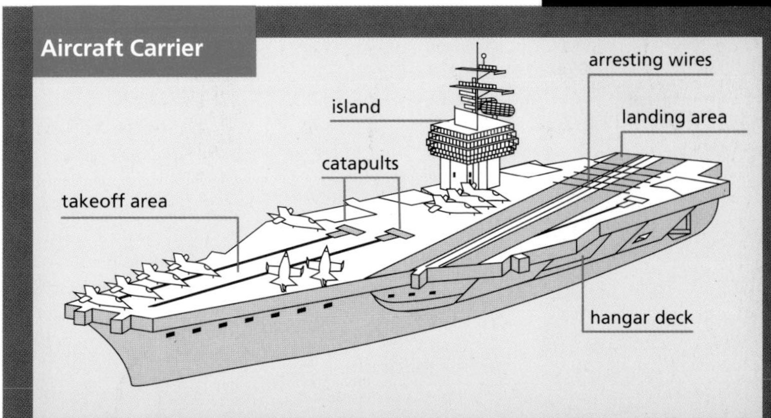

A ship designed to launch and recover aircraft. Aircraft may take off from a carrier's deck under their own power or may be assisted by catapults located on the deck. The angled deck allows airplanes to take off while others land. Carrier aircraft attack enemy targets, defend fleets of ships, and conduct reconnaissance and antisubmarine operations.

The aircraft carrier was introduced in 1911, when a biplane landed on and took off from a wooden platform on the U.S.S. *Pennsylvania.* Aircraft carriers played a crucial role in the Pacific during World War II.

Airframe

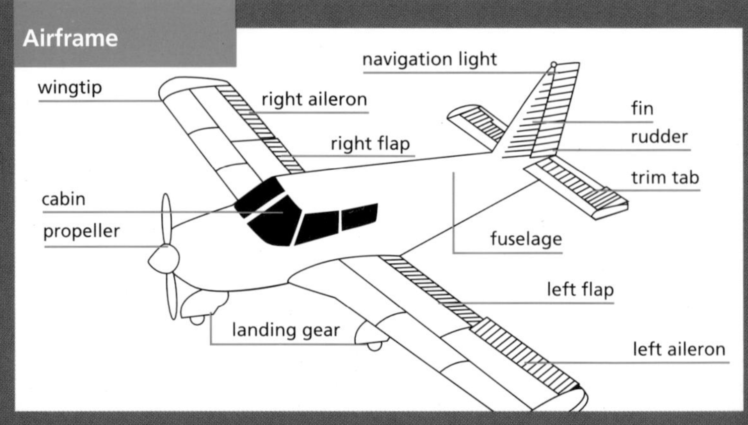

The shell of an aircraft body, including the wings, fuselage, tail assembly, and so on. The placement and size of these parts vary greatly from aircraft to aircraft. Advanced airplanes may lack some of these elements and employ new ones, such as canards, winglets, and vertical stabilizers. Still, most airplanes are equipped with the features shown here. An airplane's engines, avionics, fuel, and cargo are not considered part of the airframe.

gine. The afterburner injects fuel into the hot exhaust gases from the engine and burns the fuel to give the aircraft additional thrust.

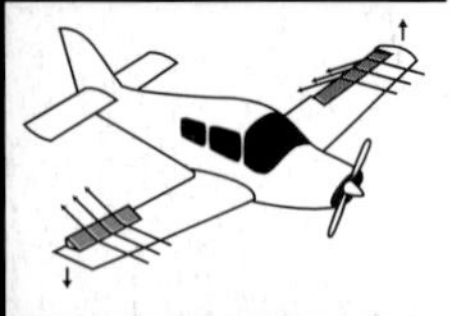
ailerons

ailerons Movable portions of airplane wings usually located at the trailing edge near the wingtips. Their function is to impart a rolling motion to the aircraft, making banking possible. They replaced the wing warping used on the Wright aircraft.

Airbus

Airbus Industrie A European aircraft manufacturers' consortium, formed in 1970, that competes with American manufacturers of large-capacity, wide-body aircraft. Airbus jets commanded 35 percent of the world market in 1990. Member companies include Aérospatiale (France), Deutsche Airbus GmbH (Germany), British Aerospace PLC (Great Britain), and Construcciones Aeronauticas SA, or CASA (Spain). The consortium is headquartered in Toulouse, France.

Air Commerce Act An act of Congress passed in 1926 that established minimum aircraft safety standards and air traffic rules. It called for the registration of aircraft, the rating of airlines and of the fitness of aircraft, and the examination of pilots, navigation facilities, and aviation schools.

Aircraft Transport and Travel (AT&T) Operating in Britain from 1916 through 1920, the first air transport company in the world to offer regular daily international passenger flights.

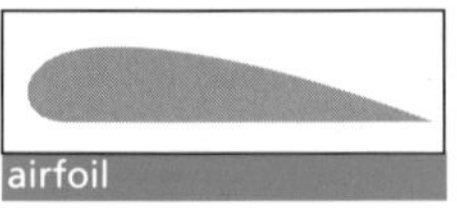
airfoil

airfoil A streamlined body, such as a wing or a propeller blade, designed to provide lift or thrust when in motion relative to the surrounding air.

Air France The airline formed in 1933 from the union of five airlines. Air France initially served Europe, the Far East, and West Africa.

Air Mail Act A congressional bill introduced by Clyde Kelly, representative of Pennsylvania, and passed in 1925. The Kelly (or Air Mail) Act took the air delivery of mail out of the exclusive domain of the U.S. Post Office and opened it up to private contractors. In time these airlines transported passengers as well as mail, ushering in the modern U.S. airline industry.

airmail The transport of letters and packages by aircraft. Mail was sometimes carried by early balloonists and air pilots as a stunt until 1911, when the first official U.S. airmail was flown between Garden City and Jamaica, New York. The U.S. Army operated the first regular airmail service from May to August 1918, when the post office took over the job.

airscrew A propeller designed to operate in air (as opposed to a marine screw, which is a propeller designed to operate in water).

airship *See* **dirigible.**

air-to-air missile A missile intended to be launched by one aircraft against another.

air-to-surface missile A missile intended to be launched by an aircraft against a ground target.

air traffic controller A person whose job is to communicate with pilots of aircraft in flight, making certain that planes follow planned routes and do not collide in midair. Using radar and computers, an air traffic controller regulates the flow of air traffic.

Air Transport Auxiliary (ATA) A British organization of civilian pilots—both men and women—whose job was to ferry aircraft to and from air bases during World War II.

Albatros

Albatros A German aircraft manufacturer founded in 1910. Albatros played an important role in World War I, producing the D.III "vee strutter" and other fighter planes as well as floatplanes and reconnaissance aircraft. A new company established in 1923, also called Albatros, built mostly training planes until it merged with Focke-Wulf in 1931.

Alcock, Captain John (1892–1919) British aviator who with fellow Briton Arthur Whitten-Brown made the first nonstop transatlantic flight. Alcock joined the Royal Naval Air Service at the beginning of World War I, took part in the bombing of Istanbul, and was captured by the Turks. After the war he became a test pilot for Vickers Aircraft, a company involved in the competition for the £10,000 prize offered by the London *Daily Mail* for the first successful transatlantic flight. He and Whitten-Brown won the prize.

Allies On September 5, 1914, Russia, France, and Great Britain signed an agreement stipulating that none of them would make a separate peace with the Central Powers, and that the three nations should henceforth be known as the Allied Powers, or simply the Allies. The United States joined them in 1917. The term is also used to designate the alliance of France, Britain, the United States, and several other countries during World War II.

altimeter An instrument that tells the pilot how far above land or sea the aircraft is flying. Some altimeters use differences in air pressure to gauge altitude, whereas others use radio waves bounced from the airplane to the earth and back to the plane.

angle of attack The acute angle between the direction of the relative wind and the position of the wings. In conventional aircraft, pilots avoid too high an angle of attack, which results in a loss of lift. *See also* **lift.**

area bombing A bombing strategy that targets entire towns or large city areas rather than a specific target.

attitude The position of an aircraft's axes relative to a particular reference point, such as the horizon or the location of another airplane.

autogyro (also spelled *autogiro*)An aircraft powered by a standard propeller for forward motion and a freespinning rotor for lift. Also called a gyroplane, the autogyro used the fuselage and propeller of a conventional airplane, but added a large rotor on top instead of conventional wings. Autogyros could take off on very short runways and are considered the forerunner of the modern helicopter.

autopilot A device for automatically steering and stabilizing aircraft.

AVCO (Aviation Corporation) A holding company organized in 1929 from the merger of the Robertson Aircraft Corporation and Colonial Air Transport. The following year AVCO was renamed American Airways, Inc. In 1934 the company expanded, acquiring U.S. routes coast-to-coast, and was reincorporated as American Airlines.

aviation The operation of heavier-than-air aircraft (from the Latin work *avis,* meaning "bird").

avionics The electronics used in aviation. Voice communication by radio was one of the first developments in avionics. Today, advances in computers, fiber optics, and other technologies are affecting avionics used in both military and general-aviation aircraft—and are used to upgrade old aircraft.

Avro

Avro A British aircraft manufacturer founded by A. V. Roe in 1908. Its famous 504 biplane, produced from 1913 until 1933, was used as a bomber and fighter in World War I. Other successes included the Lancaster heavy bomber, used in World War II, and the Vulcan delta-wing jet bomber. In the early 1960s Avro was merged into Hawker Siddeley.

AWACS (airborne warning and control system) A system that combines the capability of detecting aircraft (airborne early warning, or AEW) with the capability or facilities to respond to such detection. Uses of AWACS aircraft range from providing room for a commander and staff to develop operations while in the air, to directing interceptors toward advancing enemy aircraft.

B

balloons Lighter-than-air aircraft made up of a large bag (an "envelope") filled with gas that is lighter than the

surrounding air. Hot air is the most practical gas for amateur ballooning, but it provides a much weaker lift than either hydrogen (which is highly flammable) or helium (which is very expensive). In 1783 the Montgolfier brothers launched the first large-scale hot-air balloon, and J. A. C. Charles launched the first hydrogen balloon. Helium was used as early as the 1920s, and became the preferred gas in the late 1930s, following the tragic *Hindenburg* accident. Balloons have been used for scientific research, military observation and defense, and sport and recreation.

ball turret A spherical single-gunner station attached to the underside of a bomber. Equipped with twin 50-caliber machine guns, the ball turret could rotate in any direction required by using two overhead levers. The ball turret was added to the B-17 after raids over Europe proved that more defensive armament was needed on the bombers.

barnstormers Stunt fliers whose traveling exhibitions promoted interest in aviation. After World War I, the surplus of airplanes and shortage of jobs stimulated the growth of barnstorming by former fighter pilots. Stunts included wing walking, plane-to-plane transfers, flying under bridges, loops, and last-second pull-outs from dives.

Battle of Britain, The A battle waged from July 10 to October 31, 1940, by Britain's Royal Air Force against the German Luftwaffe. The Luftwaffe conducted air attacks on RAF airfields and bombed London. The RAF rebuilt its airfields, however, and wore down the enemy. Both sides suffered heavy losses of aircraft, pilots, and crews before the Germans switched to night bombing raids.

Lincoln Beachey

Beachey, Lincoln (1887–1915) Beachey was the preeminent stunt pilot in the period before World War I. He traveled throughout the United States staging demonstrations to promote Glenn Curtiss aircraft. He died when he failed to pull out of a dive over Oakland Bay, California.

Beech

Beech The Beech Aircraft Corporation is one of the three prominent American manufacturers of lightplanes, or "Beechcraft," as it calls them. The company was founded in 1932 by Walter and Olive Ann Beech, a husband-wife team. Known for its "Staggerwing" cabin biplanes and Bonanza monoplanes, Beech today focuses on constructing Beechcraft for business use.

Beech, Walter (1891–1950) Founder, in 1924, of the Travel Air Manufacturing Company in Wichita, Kansas, which built sporting and business airplanes. In 1932 he organized what would become the Beech Aircraft Corporation, which manufactured planes for World War II and, later, for peacetime aviation. Beech remained the company's president and chairman for the rest of his life.

Bell One of the world's largest manufacturers of helicopters, founded in 1935 by Lawrence Bell and known today as Bell Helicopter Textron. Among its designs are the P-59 Airacomet, the first U.S. jet aircraft; the X-1, the first aircraft to break the sound barrier; and the Bell 47, once the world's most widely used helicopter.

Bell, Alexander Graham (1847–1922) Scottish-born U.S. scientist. He invented the telephone before he was 30 years old. His varied interests included boats and aircraft. Bell founded the Aerial Experiment Association in 1908 to explore heavier-than-air flight.

Bendix Trophy Race Established in 1931 by Vincent Bendix, head of Bendix Aviation, this was an annual competition for the shortest time on a U.S. transcontinental air race.

biplane An airplane with two main wings.

Bishop, William ("Billy") Avery (1894–1956) Canadian ace who shot down 72 German aircraft during World War I. Bishop went overseas in the cavalry but transferred to the Royal Flying Corps in 1915. He was appointed to the staff of the British Air Ministry in 1918, and in this capacity helped to form the Royal Canadian Air Force.

blind flying The ability to maintain control of an aircraft without needing to see outside the cockpit.

Bell 47

Alexander Graham Bell

biplane

Blitz Derived from the German word *blitzkrieg* ("lightning war"). The Blitz was the period from September 1940 until the spring of 1941, during which the German Luftwaffe conducted night bombings of London, England, and surrounding areas. Although the raids were long and the bombings heavy, Britain industry remained operative. The Blitz ended with the shorter nights of spring and the transfer of German bombers to the eastern front.

blitzkrieg A German term (meaning "lightning war") used in World War II to describe the German method of using swift armor attacks supported by aircraft.

Boeing Airplane Company An airplane manufacturing company formed in 1917 south of Seattle, Washington, by William E. Boeing, renamed from the Pacific Aero Products Company. Known since 1961 as the Boeing Company, it designs and produces military and commercial airplanes, helicopters, space vehicles, missiles, and railway cars.

Boelcke, Oswald (1891–1916) An able German aviator who was the teacher of Baron Manfred von Richthofen and led a squadron that included Richthofen. He trained the famous "flying circus" that was taken over by Richthofen. He also originated the practice of forming fighter attack squadrons instead of limiting them to escort duty. When Boelcke was killed, Richthofen replaced him as squadron leader.

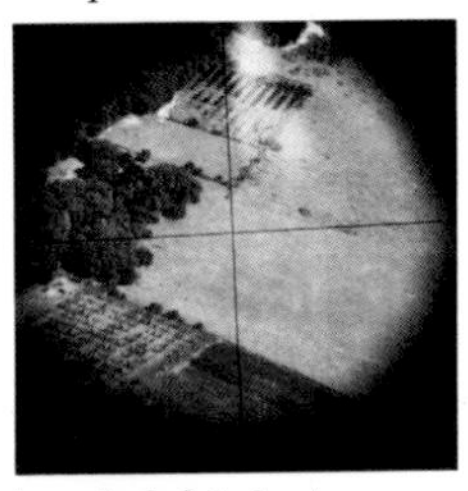

bombsight An instrument that tells the crew the appropriate time to release bombs in order to hit a target. Sophisticated mechanical and radar bombsights were developed during World War II.

Bong, Richard (1920–1945) A U.S. ace in World War II. From 1942 through 1944 Bong shot down 40 enemy aircraft, three-fourths of them fighters. He was awarded the Medal of Honor in late 1944 and died the following year in a flying accident while a test pilot for Lockheed.

boundary layer A very thin sheet of air surrounding the surface of a wing in flight.

box kite A rectangularly shaped kite invented by Lawrence Hargrave, an Australian, in the 1890s. Hargrave was lifted 16

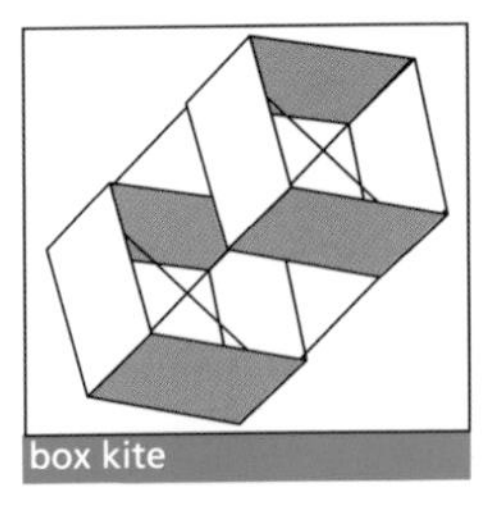

box kite

feet off the ground in 1894 by four box kites he had built. Box kites fly on one edge and do not require a tail. The Wright brothers' first flying machine, built in 1899, was a biplane kite based on the box kite's construction.

Bristol

Bristol One of the oldest manufacturers of aircraft and aircraft engines in Great Britain. Founded in 1910, the company produced such planes as the F.2B, or "Brisfit" fighter, and the Britannia transport. In 1966 the engine department (newly independent) became part of Rolls-Royce. Bristol Aircraft became part of the British Aircraft Corporation, and Westland took over helicopter operations.

British Aerospace A British aircraft manufacturer that joined Airbus Industrie in 1979 with a 20 percent interest in the consortium. Although aircraft produced by Airbus are assembled in Toulouse, France, British Aerospace is responsible for manufacturing certain parts, including major portions of the wings.

Brown, Walter Folger (1869–1961) U.S. postmaster general under Herbert Hoover from 1929 to 1933. Brown gained nearly unlimited power over the airlines in 1930 when he spearheaded the McNary-Watres Act through Congress. The provisions of the act allowed him, almost singlehandedly, to redraw the country's airline map and to reorganize the airlines that would fly it.

Commander Richard Byrd

Byrd, Commander Richard (1888–1957) U.S. naval officer, pioneer aviator, and polar explorer. Commander Byrd learned to fly at the U.S. Naval Air Station at Pensacola, Florida, served in World War I as a navy pilot, and later worked on navigational aids for airplanes and dirigibles built for transatlantic crossings. He flew as navigator in a Fokker trimotor airplane over the North Pole in 1926.

camber The arching curve of an aircraft wing from the front (or "leading") edge to the rear (or "trailing") edge.

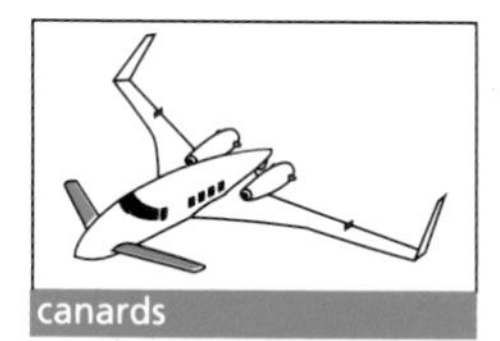
canards

canards Horizontal stabilizers located in front of the wings of an aircraft. Canards may be used to counteract the nose-down tendency of airplanes or to improve maneuverability. They have appeared on experimental craft throughout the history of aviation.

cannon An aircraft-mounted gun, twenty millimeters or larger, with a high rate of fire.

cantilever wing A wing supported only where it is attached at the fuselage, without other supports or braces.

Count Gianni Caproni

Caproni, Count Gianni (1886–1957) Head of one of the largest manufacturers of airplanes in Italy. Caproni founded the Caproni Company in 1911, one year after his first airplane was flown. He designed bombers that were produced in France and Italy during World War I. For this work he was made a count in 1940 by the king of Italy.

CATIA (computer-aided three-dimensional interactive application) A computer system that allows an aircraft to be designed on screen in three dimensions. CATIA increases efficiency, decreases costly errors, and allows many workers, from tool designers to assemblers, instant access to constantly updated information. CATIA is a key factor in the development and construction of the Boeing 777.

Cayley, George (1773–1857) British scientist, founder of the science of aerodynamics. Cayley discarded the ornithopter idea and devoted his attention to the fixed-wing aircraft, the helicopter, and the airship. He published extensive research on wing, rudder, and elevator structures that was of considerable value to later researchers. He built the first human-carrying glider in 1853.

Central Powers Germany, Austria-Hungary, and Turkey, allied together in World War I.

Cessna One of the world's largest manufacturers of light aircraft, based in Wichita, Kansas. Founded in

1927 by Clyde Cessna, the company established its reputation before World War II with its Airmaster monoplanes. Its primary wartime plane was the AT-17/UC-78, used for transport and training. Cessna continues to focus on producing jet and turboprop cargo and business airplanes.

Cessna, Clyde (1879–1954) A U.S. flier and aircraft manufacturer whose monoplane designs made him very successful. Cessna decided to become an aviator when he saw a flying circus in Oklahoma. He flew his first plane in 1911, and by 1917 had built one. In time, his Cessna Aircraft Company grew to produce more than eight thousand planes per year.

chandelle A flight maneuver in which the plane makes an abrupt 180° climbing turn just short of a stall and uses the momentum of the airplane to achieve a higher rate of climb than is normally possible.

Chanute, Octave (1832–1910) U.S. pioneer in glider design. With an excellent reputation as a civil engineer, Chanute turned his structural skills to glider planes. His gliders used vertical struts and diagonal bracing wires, and were so stable that they made two thousand flights without an accident. He was in constant correspondence with the Wright brothers at Kitty Hawk in 1901 and 1902.

J. A. C. Charles

Charles, J. A. C. (1746–1823) A French inventor and physicist who in 1783, with brothers Nicolas and Anne-Jean Robert, built the first hydrogen balloon. He and Nicolas became the first people to ascend in a hydrogen balloon when they launched their creation that same year.

Claire Chennault

Chennault, Claire (1890–1958) A U.S. major general who served more than 20 years with the U.S. Army and Army Air Corps. Chennault created the American Volunteer Group (better known as the Flying Tigers), whose first mission was to fight Japanese air superiority in 1941 and 1942 over China.

China Clipper (M-130) A four-engine flying boat, with 32 seats, built by the Martin company for Pan Am in 1935. The China Clipper carried mail, and later passengers, from Alameda, California, to Manila, the Philippines, via Hawaii, Midway, Wake, and Guam.

Civil Aeronautics Act of 1938 A congressional act that created the Civil Aeronautics Authority to license and regulate the aeronautics industry. The act was initiated after a 1934 government scandal concerning mail contracts. The Civil Aeronautics Authority was joined in 1940 by the Civil Aeronautics Board, which oversaw airline economic and safety issues.

Cochran, Jacqueline (Jackie) (1910–1980) U.S. pilot who held more speed, distance, and altitude records than any other flier during her career. She had been a beauty shop operator and had founded her own cosmetics firm—and learned to fly partly to promote it. She and Amelia Earhart were the first women to enter the Bendix Transcontinental Air Race, and Cochran won it in 1938. She was the first woman to fly faster than the speed of sound and the first woman to fly at twice the speed of sound. After retirement

Jacqueline Cochran

as a colonel from the Air Force Reserve in 1970, she served as a consultant to the National Aeronautics and Space Administration (NASA).

combined-cycle engine An engine design that would combine the functions of many engine types currently in use.

composite A material made from two separate and distinct components. Composites are often resistant to high temperature and are immune to corrosion, all desirable properties in aircraft manufacture. Although new composites are continually being developed, the oldest and most common is fiberglass, which was first used in the airline industry in the Boeing 707.

computer-aided design (CAD) A method of designing an aircraft part, or even an entire plane, using a computer screen instead of a traditional drafting board.

Consolidated An aircraft manufacturing firm established by Major Reuben H. Fleet in 1923 and originally based in Buffalo, New York. The company moved to San Diego, California, in 1935, and in 1943 became Consolidated-Vultee, which in 1944 produced more aircraft than any other company in the world. By 1953 it was known as Convair and then became Convair Aerospace Division of General Dynamics, which is now part of Lockheed.

contact flying The navigation of an aircraft in conditions of good visibility by directly observing landmarks on the ground instead of by using aircraft instruments or air traffic control guidance. The Federal Aviation Administration maintains Visual Flight Rules, which impose minimum altitudes and weather conditions on aircraft engaged in contact flying.

contrail Short for "condensation trail" (also called "vapor trail"), referring to a visible trail of condensed water vapor or ice particles that an aircraft leaves behind when flying at high altitude.

Cornu, Paul (1881–1944) The French engineer who designed and built the first helicopter to complete a free flight—20 seconds long—with a person on board (1907). (Before

Paul Cornu

this success, another helicopter design had lifted off the ground but was held in place by people standing on the ground.) Cornu's helicopter was later abandoned as impractical.

Corrigan, Douglas "Wrong Way" (1907–) After working as a welder on the *Spirit of St. Louis,* the craft that Ryan Airlines was building for Charles Lindberg, Douglas Corrigan became an enthusiastic pilot. He completed two round-trip flights across the United States in 1937, but he is best known for his July 1938 flight to Ireland. Lacking official permission to fly the Atlantic, he flew his Curtiss Robin, without a radio, to Ireland. He covered 3,000 miles in 28 hours, 13 minutes—then claimed that he had intended to fly across the United States but that a compass error had sent him in the opposite direction. His flight, and his sense of humor, brought him instant fame, a role in the 1939 movie *The Flying Irishman,* and his nickname—Wrong-Way Corrigan.

Crossfield, A. Scott (1921–) U.S. pilot and engineer who helped test and develop the X-15, a plane that was capable of reaching speeds beyond Mach 5 and which many experts feel is responsible for paving the way for piloted spaceflight. Crossfield piloted the X-15 during its early test flights, including its first unpowered flight in 1959, when it was released from a B-52 from 38,000 feet and glided safely to earth. In later flights, Crossfield engaged the plane's powerful engines and pushed it to Mach 2.4. Despite the dangers of flying the X-15, Crossfield's closest brush with death ironically came during a ground test of X-15 #3. He was in the cockpit when the engine exploded and destroyed the airplane aft of the cockpit.

Curtiss Founded in 1909 by Glenn Curtiss and Augustus Herring as the Curtiss-Herring Company, Curtiss was one of the oldest builders of aircraft in the United States. Its most famous planes include the Jenny and the NC-4 and other flying boats. In 1929 Curtiss merged with the Wright Aeronautical Company to form the Curtiss-Wright Company.

Curtiss, Glenn (1878–1930) Glenn Curtiss began his career as a mechanic who raced motorcycles. His design and manufacture of lightweight, high-speed gasoline engines led to work making engines for airships. He caught the attention of Alexander Graham Bell, who invited him to become part of the Aerial Experiment Association. Curtiss was a strong competitor of the Wrights from 1908 to 1914. His airplanes differed from the Wright planes in the use of ailerons rather than wing warping to control banking, and in the use of wheeled landing gear rather than skids. He was the first builder of seaplanes in the United States, and had the first contract to build U.S. Navy planes. His best-known plane was the Jenny, widely used as a trainer in World War I and later by barnstormers.

Curtiss-Wright The company formed in 1929 with the merger of the Curtiss Aeroplane and Engine Company and the Wright Corporation. Curtiss-Wright produced a wide range of military aircraft and aircraft engines, including the Hawk family of fighters and the Cyclone series of engines. In 1947, the company closed its airplane division.

D

Dash Eighty The prototype of the Boeing 707, first flown on July 15, 1954. The Dash Eighty had four turbojet engines, a wingspan of 130 feet, and a takeoff weight of up to 190,000 pounds. It was the prototype for the first of a family of jet transport airliners.

Dassault-Breguet Founded by Marcel Bloch shortly after World War II, the Société des Avions Marcel Dassault became the only French manufacturer of jet fighters. Its greatest success was the Mirage, which could fly at Mach 2 speeds. In 1971 the company merged with Société Louis Breguet to form Dassault-Breguet, now called Dassault Aviation.

de Bothezat, George A. Russian-born engineer who in 1922 unveiled the first successful experimental helicopter, at what is now Wright Field in Dayton, Ohio. Not intended for commercial or private use, de Bothezat's helicopter nevertheless made approximately 50 flights, proving that with further development the machine had tremendous potential.

de Havilland, Geoffrey (1882–1965) Founder of the de Havilland Aircraft Company in 1920 in Great Britain. The company produced the popular Moth and other civil airplanes until World War II, when it mass-produced the Mos-

Geoffrey de Havilland

quito. De Havilland also developed the first jet airliner, the Comet. Its engine division, which produced the Gipsy, the Goblin, and the Ghost, was eventually absorbed by Rolls-Royce.

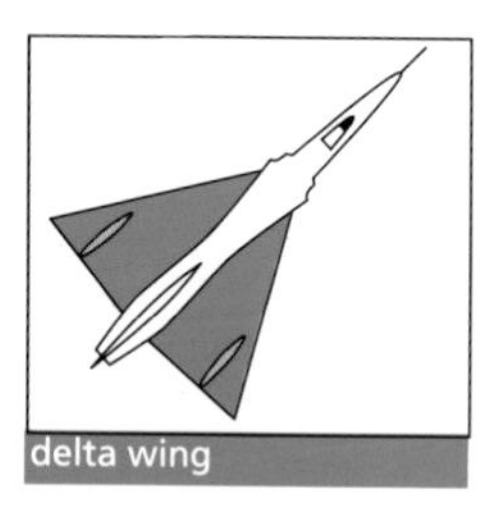
delta wing

delta wing Derived from the shape of the Greek letter *delta,* a delta-wing design merges the wing and tail of an aircraft into one large flying surface. The shape of the wing resembles an isosceles triangle and provides greater aerodynamic advantage in supersonic flight.

Deperdussin Founded by Armand Deperdussin in 1910, this French aircraft manufacturer became famous for its racing monoplanes with advanced construction. Deperdussin also manufactured the first planes with monocoque ("single-shelled") fuselages, which used multiple layers of thin tulip wood. The company was taken over by SPAD after Deperdussin was convicted of fraud.

de Saint Exupéry, Antoine (1900–1944) Award-winning novelist, professional mail pilot, commercial test pilot, author of *The Little Prince*, war correspondent, and World War II veteran. He combined his careers as pilot and author by keeping a notebook with him on every flight and even occasionally using a Dictaphone in the cockpit. *Night Flight* describes his experiences as an early airmail pilot flying at night and through storms at sea; *Wind, Sand, and Stars* recounts his crash landing in the Libyan desert. In *Flight to Arras* he reflects on the defeat of France and his experience as a pilot in World War II.

Deutsche Luft Hansa

Deutsche Lufthansa A German company formed in 1926 from the merger of two companies. Deutsche Luft Hansa was renamed Deutsche Lufthansa in 1930.

Dewoitine A French aircraft manufacturer founded in the early 1920s by Emile Dewoitine. Dewoitine built transports and fighters, most notably a series of fighters beginning with the D500 in 1932. In time Dewoitine formed the Société Aéronautique Française, which was nationalized as SNCAM in 1936.

digital flight-control system (DFCS) An experimental "smart" flight-control system, or computer, through which a pilot flies an aircraft. A DFCS can even be programmed to make sure that all maneuvers can be performed within the aircraft's structural limits, preventing, for example, the wings being ripped off while pulling out of a dive.

dogfight A fight in aerial warfare between two or more fighter planes usually maneuvering at close quarters.

Doolittle, Lieutenant Colonel James (1896–1993) A World War I aviator and flight instructor, Doolittle earned a doctorate in advanced engineering from M.I.T. after the war and led the effort to develop blind-flying instruments. In 1929, in a fully hooded cockpit, he made the first completely blind flight in history, taking off, flying a predetermined course, and landing, all by instruments alone. Doolittle set the speed record for landplanes in

1932. Four months after the Japanese attack on Pearl Harbor, he led a strike force of 16 B-25 bombers in a raid on Tokyo.

Dornier After beginning his career in aviation with Zeppelin Werk Lindau in World War I, Claude Dornier reorganized the company as Dornier Metallbauten in 1922. Its flying boats were widely used for military and mail-route services, but it became best known during World War II, when it produced bombers and night fighters for the Luftwaffe.

Douglas, Donald (1892–1981) An American aircraft designer who organized the Douglas Aircraft Company in 1920. His World Cruiser biplanes completed the first around-the-world flight in 1924, and his DC series of commercial planes (including the DC-3 and the DC-9 and DC-10 jet transports) proved successful. In 1967 the company became a division of the McDonnell Douglas Corporation.

Dowding, Sir Hugh (1882–1970) The British air chief marshal who was largely responsible for the defeat of the Luftwaffe in the 1940 Battle of Britain. Named chief of the Fighter Command in 1936, Dowding led the command with strategic and tactical skill, which led to Britain's victory, despite its being outnumbered by the Germans.

drag The force acting on an object to hinder forward motion. When an object moves through air, drag is produced.

drift The tendency of an aircraft, bomb, or missile to stray off course due to the effect of the wind.

Earhart, Amelia (1897–1937) U.S. aviator and the first woman to fly alone across the Atlantic Ocean. Amelia Earhart was a military nurse in Canada during World War I, and later a social worker in Boston. She made her solo flight across the Atlantic in 1932. She actively worked for the development of commercial aviation and for the end of male domination in the new

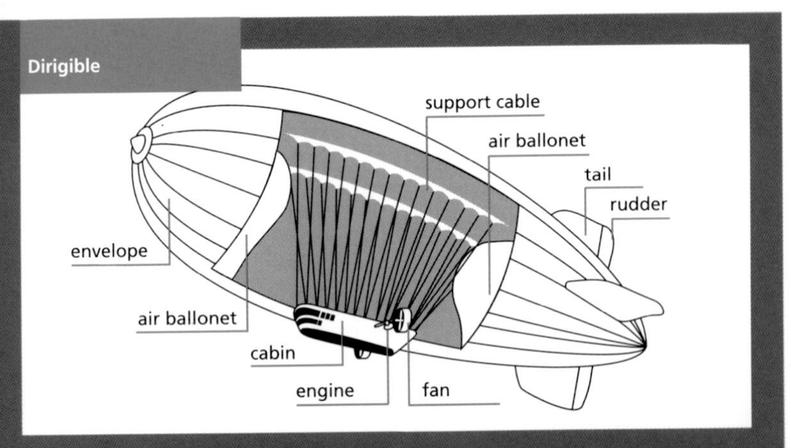

A steerable (*dirigible* means "steerable"). Rigid and nonrigid dirigibles filled several roles—reconnaissance, surveillance, bombing, and passenger transport—though after the *Hindenburg* disaster in 1937, the fragility of the rigid type of airship and the flammability of hydrogen ended its era. New and safer airships are currently making a comeback.

field. In 1937 she set out, with Fred Noonan as navigator, to fly around the world. They had completed more than two-thirds of the trip when their plane disappeared over the central Pacific.

Eindecker A German word meaning "one-winger," the name given to the Fokker single-seat monoplane. This airplane became a deadly fighter, equipped with a machine gun with an interrupter to avoid propeller damage. Eindeckers caused the "Fokker Scourge" from October 1915 to May 1916.

ejection seat A means of bailing out of an airplane in an emergency. A rocket motor shoots the pilot and seat out of the airplane. Once the pilot is clear, an automatic parachute opens.

elevator A movable airfoil, usually attached to the tail of a plane. Its function is to cause an airplane to dive or to climb.

elevon Control surfaces that combine the tasks of ailerons and elevators, aiding in lateral control and vertical movement.

empennage The complete tail unit of an aircraft. It gets its name from the French word for the feathers at the end of an arrow, which it resembles.

engine cowling A removable metal covering that houses the engine.

engine pod A nacelle, or enclosed shelter, on an aircraft that houses one or more jet engines.

"Little Boy"

Enola Gay The specially equipped American B-29 that dropped "Little Boy," an atomic bomb, on Hiroshima, Japan, on August 6, 1945. The bomb devastated a 4.7-mile area of the city, killing more than 66,000 Japanese.

Experimental Aircraft Association (EAA) An organization founded in 1953 to support home-built aircraft development and interest. Local EAA chapters nationwide provide encouragement and assistance to individual home-builders. Each year in Oshkosh, Wisconsin, the EAA sponsors the world's largest international airplane convention, featuring displays, design forums, builders' meetings, and daily air shows.

Fairey The British aircraft manufacturer formed in 1915. Fairey produced predominantly military aircraft, including the IIIF seaplane (1926) and a variety of fighters and bombers. After World War II Fairey built antisubmarine, airborne early warning, and research delta-wing aircraft, as well as helicopters.

Fairgrave, Phoebe (1903–) When licensing of pilots began in the United States in 1926, Phoebe Fairgrave was one of the first to be licensed. In the 193C she organized a group (women fliers who barn stormed the country to urge communities to paint the name of their town or city in large white letters on a rooftop to aid pilots in navigation. The first woman to hold a government aviation post, she was appointed by President Franklin D. Roosevelt as technical adviser to the National Advisory Committee for Aeronautics (NACA).

fairing A secondary structure fitted onto an aircraft, smoothing joins between surfaces to transform the fuselage into a sleek form. Fairings do not add to the strength of the aircraft.

Farman One of the largest French aircraft manufacturers before World War II, Avions H. and M. Farman was founded in 1912 by two French brothers. Farman produced a variety of military aircraft as well as civilian planes for passenger and mail traffic. In 1936 the company was merged with Hanriot and became the Société Nationale de Constructions Aéronautiques du Centre (SNCAC).

Farman, Henri (1874–1958) **and Maurice** (1877–1964) French fliers and aircraft designers and manufacturers. In 1908 they made the first circular flight of more than one kilometer. In 1917 they introduced the Goliath, a large passenger plane, which made regular flights between Paris and London after 1919.

fin The fixed vertical part of an aircraft's tail assembly.

flak The fire from antiaircraft guns located on the ground or on ships and fired in defense against attack from the air. Antiaircraft devices first appeared in World War I, but great improvements in range and accuracy were made between the world wars. Beginning in the 1950s antiaircraft efforts also included radar-and-computer-operated automatic cannon and ground-to-air rockets and missiles.

flap An extension of the wing used to modify its lift characteristics. In takeoff and approach to landing they supply greater lift than drag. During flight or after touchdown they can increase drag for quick deceleration.

fly-by-light control system An aircraft control system that uses optical fibers to transmit signals at the speed of light, using light from a laser. This system replaces the heavy wiring of a fly-by-wire system with very light fiber optics, which can also withstand electromagnetic radiation, giving military aircraft an important advantage.

fly-by-wire control system An aircraft control system that uses wires and other electromechanical devices, including a digital computer, to transmit electrical signals. Fly-by-wire systems were first used on spacecraft. Military fighters, the Concorde, and some commercial airliners also use fly-by-wire systems.

flying boats Seaplanes built with an integrated, boatlike hull rather than separate pontoons.

flying circus World War I nickname for a large formation of military aircraft, or a stunt-flying show performed after World War I, often led by a military pilot who had returned from the war. Flying circuses would perform daring feats such as airplane maneuvers, parachute jumping, and standing or walking on the wings of a flying airplane. Pilots of these craft, also known as barnstormers, earned money by giving rides to observers.

Flying Tigers The common name of the American Volunteer Group (AVG), an organization of U.S. pilots founded by General Claire Chennault in 1941. Despite their obsolete airplanes and fuel shortages, the Flying Tigers inflicted considerable damage to Japanese forces in Burma and China, using surprise, mobility, and unconventional tactics. The AVG was disbanded on July 4, 1942.

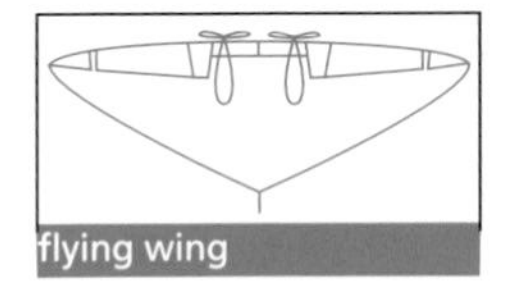
flying wing

flying wing An aircraft that has only wing surfaces, eliminating the drag generated by a fuselage. Flying wings were first designed in the 1940s but were not mass-produced because of their instability. Current computerized control systems can compensate for this limitation, however, and craft such as the Northrop B-2 now employ this design.

Focke, Heinrich A German aircraft manufacturer who in 1937 completed the best helicopter up until that time, the Focke-Achgelis windmill. Focke's helicopter was the first to be fully controlled, could be held aloft for more than 1 hour and

20 minutes, and reached a height of 7,900 feet.

Fokker A European aircraft manufacturer, founded by Anthony H. G. Fokker just before World War I. Fokker produced many fighters used in both world wars. It also built successful passenger planes, including the aircraft flown by Richard Byrd over the North Pole and by Amelia Earhart across the Atlantic. Fokker is now part of Zentralgesellschaft VFW-Fokker mbH in Amsterdam, the Netherlands.

Anthony Fokker

Fokker, Anthony (1890–1939) Dutch airman and pioneer aircraft manufacturer, who offered his designs to both the Allies and the German High Command. The Allies turned him down, and he produced more than 40 types of aircraft for the Germans in World War I. He introduced the gear system that synchronized the propeller and the machine gun so that the bullets passed through the propeller area without striking the blades. After the war he sold aircraft to the U.S. military and opened a second factory in New Jersey.

Fokker Scourge The period from October 1915 until May 1916, during which Germany's Fokker planes virtually controlled the skies in World War I, outmaneuvering and outgunning the planes of the Allies.

René Paul Fonck

Fonck, René Paul (1894–1953) The leading World War I ace on the Allied side, with 75 German planes shot down in two years, without receiving a scratch himself. He was the first to try for the coveted Orteig Prize, seven years after it was first offered. However, his plane crashed on takeoff, leaving the prize for Charles Lindbergh to claim.

Ford An American automobile manufacturer founded by Henry Ford in 1903. The Stout Metal Airplane Division of Ford produced the Ford Tri-motor monoplane airliner from 1926 to 1933, a very popular aircraft used on routes throughout the United States and in a number of other countries. Ford also manufactured B-24 Liberator bombers during World War II.

fuselage The central body portion of an airplane designed to accommodate the crew and the passengers or cargo. The word comes from a French word meaning "spindle."

G force A measure of acceleration equal to the acceleration of gravity (32.2 feet per second at sea level). Pilots must take G forces into account when they are recovering their aircraft from dives, for example, to keep from exceeding the structural limits of the aircraft.

G suits A nylon suit that fills with air to pressurize the abdomen and lower body, preventing blood accumulation below the chest. The suit helps pilots withstand high G forces.

Garros, Roland (1888–1918) French pianist and pilot, taught to fly by Santos-Dumont. In 1915 Garros fitted a plane with a machine gun that could fire through the propeller, because the propeller blades had steel deflecting plates attached to them. Garros was the first to fly across the Mediterranean, and he set an altitude record of 18,000 feet in a plane not equippid with oxygen.

glider A motorless winged aircraft, which when launched from a height descends on a path toward the ground. Upward air currents and the design of the glider can greatly increase the time of flight and the distance covered. Modern glider flights have exceeded 70 hours in flight time and 600 miles in distance.

Gloster Founded in 1915 as the Gloucestershire Aircraft Company, this British manufacturer produced 23 fighter designs for 19 air forces during the 1920s. It also produced the first British jet aircraft (1941) and the first operational Allied jet aircraft in World War II (the Meteor, in 1944). Gloster became part of Hawker Siddeley in the 1960s.

"God's eye view" The "big-picture" view of combat, projected on a pilot's head-up display (HUD), from the perspective of a point outside (such as above and behind) the aircraft.

Gordon Bennett Trophy A racing award offered by and named for James Gordon Bennett. The trophy was first offered in 1909 at an international aviation meet in Rheims, France. Glenn Curtiss won this first race, reaching a speed of 47.65 miles per hour. Claude Grahame-White and Charles Weymann won in the two succeeding years. The final race was held in France in 1920.

Göring, Hermann (1893–1946) The commander of the German Luftwaffe during World War II. Göring earned a distinguished reputation as a World War I fighter pilot. He went on to become one of Hitler's closest associates in the Third Reich. Sentenced to death at the Nuremberg war crimes trial, Göring committed suicide before his execution could be carried out.

Gossamer Albatross A human-powered aircraft designed and built by Paul MacCready, an American. In 1979 Bryan Allen pedaled the Gossamer Albatross from Kent, England, to Cape Gris-Nez, France, in the first human-propelled flight across the English Channel. This achievement won the Kremer Prize of £100,000.

Gotha Famous for its World War I bombers, which raided southern England toward the end of the war, the German company Gothaer Waggonfabrik primarily built railroad cars and diesel engines. It returned to aviation manufacturing in 1933, however, with the production of military trainers and gliders.

Claude Grahame-White

Grahame-White, Claude (1879–1959) An English pilot who was the first in that country to earn the aviator's certificate of proficiency. He entered many flying races, winning the Gordon Bennett Trophy in 1910. During World War I he supervised the construction of government planes. He later wrote about the history and development of aircraft.

Grumman A U.S. aircraft manufacturing company founded in 1929 by Leroy Grumman in New York. The company specialized in naval fighters, producing the F4F Wildcat, the F6F Hellcat, the F8F Bearcat, and other fighters during World War II. Grumman's other products included amphibians, antisubmarine aircraft, airborne-early-warning aircraft, and jet aircraft, including the F-14 Tomcat Navy fighter.

gyroscope A wheel spinning rapidly in a circular path and mounted inside a larger frame. Because of its momentum, the rotating wheel keeps a constant position, even when its supporting frame changes position. This characteristic makes a gyroscope extremely useful when incorporated into an aircraft's instrumentation. It allows the pilot to control the heading and attitude when visibility is limited. Its use in airplanes was introduced on a wide scale by Lawrence Sperry in the 1920s.

gyroscopic gunsight A type of gunsight developed in 1940 that combined a gyro and an electronic system to help achieve accurate firing of aircraft guns. The gyroscopic gunsight was a vast improvement over the ring-and-bead sight (used throughout World War I), which did not allow for the relative motion between the gunner and the target.

H

Hamilton, Charles (1885–1914) Generally considered the first true cross-country flier in the United States, Hamilton won a prize of $10,000 in 1910 for completing a flight from New York to Philadelphia and return, in 3 hours, 27 minutes of flight time.

Handley Page Founded by Frederick Handley Page in 1909, this British manufacturer specialized in large planes with multiple engines. It produced bombers during both world wars (notably the Halifax, which dropped 227,610 tons of bombs in World War II) as well as airliners for civil use. Handley Page closed its doors in 1969.

hangar An enclosed area for housing and repairing aircraft

Hawker A company established in 1920 by British aviator Harry G. Hawker following the dissolution of Sopwith Aviation, where Hawker had worked as a test pilot. Hawker died in 1921, but his company became a respected builder of combat aircraft, including the Hurricane, which achieved fame in the Battle of Britain.

heading The compass direction in which an aircraft is traveling.

heavier-than-air Of greater weight than the air displaced, as in a powered airplane.

Hegenberger, Albert (1895–1983) Navigator on the first flight from the United States to Hawaii, in 1927. Lester Maitland piloted the plane, a Fokker trimotor.

Heinkel The German Ernst Heinkel formed Heinkel Flugzeugwerke AG in 1922 and soon earned a reputation for his seaplanes. Later, Heinkel fighters and other combat aircraft, including the He 111 bomber, were used by the Luftwaffe in World War II. It produced the first jet aircraft and engine to fly, in 1939. Heinkel became part of VFW in 1965.

helicopter An aircraft powered by an engine that turns a set of overhead horizontal rotors, enabling the craft to take off and land vertically and to hover in the air. Early helicopter designs were developed by Leonardo da Vinci in the fifteenth century, but a successful machine was not developed until World War II.

helix The spiral path traced by a propeller tip of an aircraft in motion.

Henson, William Samuel (1805–1888) Collaborator, with John Stringfellow, in making far-reaching predictions on the format of modern airplanes and on intercontinental airline operations. In 1843 they published drawings that crystallized the fixed-wing monoplane configuration with main wings, fuselage, wheeled undercarriage, propeller propulsion, and a tail unit with elevator and rudder.

Herring, Augustus (1865–1926) U.S. flier and experimenter who built several Lilienthal-type gliders. He was an active and successful flying-machine experimenter in the country at the end of the nineteenth century. He became an associate of Chanute, Langley, and Curtiss.

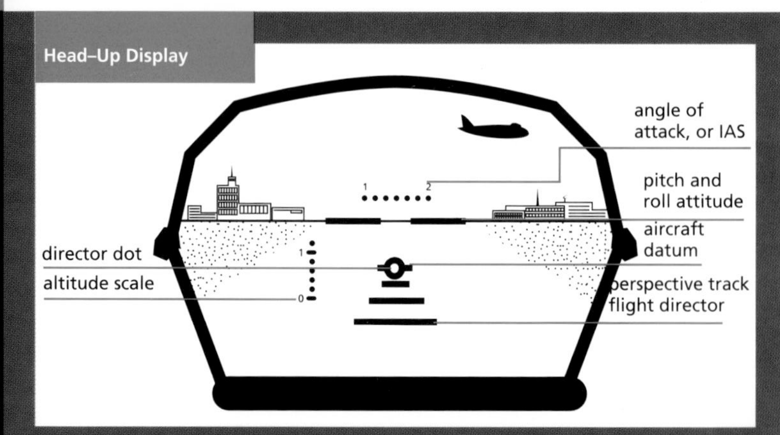

A clear screen in front of the pilot that displays information about the engine, flight control, navigation, and weapons systems while simultaneously allowing the pilot to see ahead of the plane. HUDs are in use in fighters and are now being introduced in airliners and business planes. Automakers are also developing HUDs for use in cars. These HUDs would display such data as speed, distance traveled, and warnings to check oil, gas, or brakes. In the future, they might even display traffic signs.

Hiller, Stanley, Jr. (1924–) The maker of the first coaxial helicopter. The son of an inventor, Hiller spent his childhood tinkering with engines. He founded his first company at age 16 and began developing the coaxial Hiller-copter by the time he was 19. After World War II Hiller founded Hiller Aircraft Corporation, which produced helicopters used in the Korean War.

Hindenburg A hydrogen-filled airship completed in 1936. The most luxurious passenger airship ever built, it regularly carried up to 70 passengers per trip between Friedrichshafen, Germany, and Lakehurst, New Jersey. When the *Hindenburg* burst into flames on May 6, 1937, while approaching its moorings, killing 35 passengers, the era of commercial airships came to an end.

homebuilt An aircraft constructed at home, usually from a kit. Homebuilders may take from one to seven years of their spare time and effort to complete a project. They often experiment with new designs, and their discoveries have influenced commercial aircraft manufacturing.

horsepower A standard unit of power; the rate at which work is done. One horsepower is the power needed to lift 33,000 pounds one foot in one minute, and is equal to 746 watts.

HOTAS A "hands-on throttle and stick." This control stick allows a pilot to control virtually every aircraft system at the mere touch of a button. In addition to reducing workload, the HOTAS enables a pilot to react quickly in situations that require fast decision making, without removing hands from the controls.

Howard Hughes

Hughes, Howard (1905–1976) U.S. manufacturer, aviator, and motion-picture producer. He quit school to head his father's tool company, then moved to Hollywood to produce movies, including

Hell's Angels, Scarface, and *The Outlaw.* He founded Hughes Aircraft Company, and with the profits from it, financed the Howard Hughes Medical Institute. In 1935 Hughes established the world's landplane speed record of 352 miles per hour in a plane he designed. He was the principal stockholder and guiding genius of Trans World Airlines from 1939 until 1961.

hypoxia Oxygen deprivation. This was a grave threat to the first high-altitude pilots.

I

Ilyushin A Soviet Central Design Bureau aircraft design team formed in the 1930s and directed by Sergei Vladimirovich Ilyushin. Ilyushin's team designed the Il-2 Shturmovik ground-attack aircraft and the DB-3, which was virtually the only long-range Soviet bomber of World War II. The team also designed airliners and the Il-28, the first Soviet jet bomber to be produced.

Immelmann, Max (1890–1916) German ace who pioneered the tactic of "hiding in the sun," so that his victim could not see him until it was too late. He also developed the "Immelmann turn," a maneuver in which the attacker dives past the enemy craft, pulls sharply into a vertical climb, and then rolls over, ready to dive again. When he was shot down, the government claimed he had crashed due to mechanical failure. Enraged, Fokker (the airplane's manufacturer) examined the wreckage and showed that wire stays had been severed by bullets.

Imperial Airways A British company formed in 1924, serving Europe, Africa, the Far East, and Australia. In 1940 Imperial Airways merged with British Airways to form the British Overseas Airways Corporation.

incendiary bomb A type of missile, also known as a fire bomb, containing chemicals that ignite on contact.

infrared The invisible rays just beyond the red end of the visible spectrum of light. Infrared waves reflected or given off by an aircraft can be detected by radar or other sensors.

instrument landing system (ILS) An electronic system designed to help pilots land in conditions of poor visibility. A radio system and cockpit display using horizontal and vertical lines allow the pilot to align the airplane precisely with the center of the runway. An ILS can also be linked to a plane's automatic pilot.

J

Jagdstaffel German for "hunting pack," a group of fighter planes flying in formation, protecting one another. This concept was proposed by the German flier Oswald Boelcke. First accepted by the German air force, it soon became standard practice on both sides.

Johnson, Clarence "Kelly" (1910–1990) An aeronautical engineer who joined Lockheed Corporation in 1933 and, during his more than 50 years there, played a significant role in the design and production of aircraft. His designs included fighters used in World War II, high-speed commercial and reconnaissance planes, spacecraft, and experimental craft.

Johnstone, Ralph (1886–1910) Early exhibition flier trained by Orville Wright. He was killed in an exhibition at Denver, Colorado.

jumbo jet A wide-bodied jet airliner de-

Identifying Military Aircraft

All military aircraft in the service of the United States are identified by a combination of letters and numbers, but the meaning of these codes can be confusing. In the simplest terms, letters designate a plane's role, and numbers indicate when it appeared.

Letters

Until 1962, each branch used its own set of letters to indicate the role or mission of each plane. Unfortunately, this created some confusion when, for example, T stood for trainer in the army air force and for torpedo in the navy. In 1962 the Tri-Service System was put into effect. The most common letters currently in use have the following meanings:

- **B** Bomber, as in B-52.
- **C** Cargo, as in C-130.
- **F** Fighter, as in F-86 Sabre or F-16 Fighting Falcon.
- **T** Trainer, as in T-38.

During the development of a plane, these designations indicate the phase of testing:

- **X** Experimental, as in XF-105 or X-15.
- **Y** Service Test, as in YF-111A.

Two designations were discontinued in 1962, but are still commonly seen:

- **A** Attack,as in A-4 Skyhawk.
- **P** Pursuit, as in P-51 Mustang.

Numbers

In most cases, the number assigned to an airplane indicates the order in which it was approved by the service using it. Under the new system, started in 1962, the numbering began again from 1. Today, the F-111 is an older plane than the F-16, and the B-52 is older than the B-1.

signed to carry hundreds of passengers. The Boeing 747, produced in the late 1960s, was the first jumbo jet.

Junkers The German aircraft manufacturer, founded in 1895 by Hugo Junkers, which in 1915 built the world's first all-metal aircraft. Junkers is best known for its World War II production of the Stuka dive bomber and the Ju 88 bomber. Junkers also built the first operational jet engines and a major airliner, the Ju 52 3m.

K

Kaman, Charles (1919–) An engineer and manufacturer of helicopters used in the Korean War, Kaman worked at the United Aircraft Corporation from 1940 to 1945, when he founded the Kaman Aircraft Corporation.

kamikazes Japanese pilots who volunteered for suicide attacks in the closing months of World War II. Kamikaze planes were packed full of bombs and flown directly into Allied warships, causing tremendous damage. More than 300 Allied ships were hit by kamikaze attacks during the final nine months of the war.

Kelly Bill *See* Air Mail Act.

Kevlar A composite material hard to detect by radar.

Kindelberger, James "Dutch" (1895–1962) Airplane manufacturer and executive, whose company, North American Aviation, contributed a large number of planes to the Allied cause during World War II. He is credited with introducing modern manufacturing methods into the aircraft industry. Some of the most famous planes built by North American were the P-51 Mustang and the B-25 Mitchell bomber. After the war Kindelberger's company continued to manufacture military aircraft, among them the F-86 Sabre, a jet fighter.

KLM poster

KLM Also known as Royal Dutch Airlines, a company founded in 1919. KLM was the first European airline to use DC-2s and DC-3s.

Lafayette Escadrille
Before the United States entered World War I, U.S. pilots could join the war effort only by taking an oath of allegiance to England or France, thereby forfeiting U.S. citizenship. To avoid this, some joined the French foreign legion, which required no such oath, and then were transferred into a flying group called Escadrille Américaine. Following German protests to the U.S. State Department, claiming that this was a breach of neutrality, the name was eventually changed to the Lafayette Escadrille. Since Lafayette was French, Germany could not complain. This small, elite fighter group helped gain U.S. support for the war effort, and scored many victories against the Germans.

laminar flow control
A technique for reducing the drag of an aircraft by constructing its wings so that, in high-speed flight, the boundary layer of air flows smoothly along the contours of the wing. Today's computer designs and composite materials make it possible to build an ideal wing contour with a very smooth surface or with ducts that filter the air smoothly through the wing. *See also* **boundary layer.**

Langley Aeronautical Laboratory The first research laboratory of the National Advisory Committee for Aeronautics (NACA), located at Langley Field, Virginia. Opened in 1917, by 1930 it was known as the greatest aeronautical research center in the world.

Langley, Samuel P.
(1834–1906) U.S. astronomer, physicist, and aeronautics pioneer. The third secretary of the Smithsonian Institution, Langley built the first successful unpiloted heavier-than-air flying machine, powered by a steam engine. Later he built a gasoline-powered craft, designed to be piloted and launched from a catapult. On two attempted flights, the plane plunged into the Potomac. Langley did extensive aeronautic research using a whirling arm, with tip speeds up to 70 miles per hour, in place of a wind tunnel. Many of his experimental flights were observed and photographed by Alexander Graham Bell.

Lear, William
(1902–1978) A U.S. inventor and businessman, Lear designed the first working car radio and a miniature autopilot for use on small planes. His first companies produced radio and navigational equipment for small planes. Later, he formed Lear Jet, Inc., which built its first small jet in 1963 and still specializes in private jet aircraft.

LeMay, General Curtis
(1906–1990) An officer in the U.S. Air Force whose strategic bombing techniques in the Far East during World War II helped force the Japanese to surrender. He retired in 1965 as air force chief of staff and in 1968 ran for vice president on the American Independent party ticket, with George Wallace.

Leonardo da Vinci (1452–1519) An Italian painter, sculptor, architect, and engineer. Leonardo's writings and drawings pertaining to flight were the first scientific studies of the subject. Among his design sketches of flying apparatuses are a simple helicopter, a parachute, and a propeller.

Lewis gun A machine gun with minimal recoil, invented by Isaac Newton Lewis. Lewis patented the gun in 1911, but the U.S. Army showed little interest in it. He set up a factory in England, producing more than 100,000 guns for the Allies. Lewis guns were mounted on World War I airplanes such as the Nieuport 11.

Liberty engine A U.S. aircraft engine developed in 1917, said to have been designed and built in just three months following the United States' entry into World War I. It powered aircraft such as the D.H.4.

lift When air flowing over the top surface of a wing is faster than the air running on the bottom surface, the pressure on top is lowered. The greater pressure from below propels the wing up into the low pressure area, creating lift.

lighter-than-air Of less weight than the air displaced, as a balloon filled with hot air or a gas such as helium. A lighter-than-air craft does not require power to stay aloft.

Lilienthal, Otto (1848–1896) German aeronautical researcher. He studied the flight of birds, with his brother Gustav, and published a book on bird flight in 1894. He began his experiments with ornithopters, craft with flapping wings, but soon switched to testing monoplane and biplane gliders, and was successful in achieving controlled glider flights. His extensive research on increasing the stability of aircraft was of considerable value to later experimenters.

Charles Lindbergh

Lindbergh, Charles (1902–1974) U.S. aviator, first to fly solo nonstop across the Atlantic Ocean. Lindbergh dropped out of the University of Wisconsin to enroll in a flying school, then purchased a World War I Jenny and became a stunt flier and an airmail pilot. A group of St. Louis businessmen backed him in his bid for the $25,000 transatlantic prize. In his plane, the *Spirit of St. Louis,* he successfully completed the flight in 1927. Decorated by the German government in 1938, he urged American neutrality in World War II. After being criticized for this by President Roosevelt, he resigned his Air Corps Reserve commission in 1941. He became a civilian consultant to Ford Motor Company and to United Aircraft Corporation, and in the latter capacity flew combat missions in the Pacific in World War II. In 1954 President Eisenhower appointed him a brigadier general in the U.S. Air Force Reserve.

Lockheed A major American aerospace company founded in 1926 by Allan Loughead (who changed the spelling of his name). Lockheed has produced a wide variety of combat aircraft for the U.S. military, including the F-117A stealth fighter, designed in the 1970s by Lockheed's "Skunk Works" division. The company also builds missiles, space satellites, and electronic products.

Loening, Grover (1888–1976) Often considered the first American aeronautical engineer, having received one of the first degrees in the field from

Columbia University in 1911. He was taught how to fly by Orville Wright, and after working for the Wright brothers' company, formed his own airplane firm, which evolved into Grumman Aircraft. His most famous achievements are the development of retractable landing gear and a rigid strut-bracing system.

Luftwaffe The name of the German air force since the World War II era.

Frank Luke, Jr.

Luke, Frank, Jr. (1897–1918) U.S. ace during World War I, known primarily for his expertise in bringing down German observation balloons. Luke often did not attack in broad daylight but waited until the balloons were being lowered in the evening. He usually attacked with other fliers, who acted as a diversion for the Fokkers protecting the balloons. Luke shot down a record 18 planes and balloons in 17 days. He died during a solo raid.

M

Macchi An Italian aircraft manufacturing company founded in 1912 as Nieuport-Macchi to build primarily French Nieuport aircraft. Its own designs included flying boats, which led to its concentration in marine aviation. Known later as Aeronautica Macchi, or simply Aermacchi, the company developed successful World War II fighters and postwar trainers.

MacCready, Paul (1925–) An aerodynamicist, MacCready directed a team that designed and built the first human-powered aircraft (the *Gossamer Condor*) and the first solar-powered aircraft (the *Solar Challenger*) capable of sustained flight. With degrees in physics and aeronautics, MacCready has devoted himself to the development of air and surface craft that improve air quality and conserve energy.

Machmeter An instrument that measures and indicates (with a Mach number) the speed of an aircraft relative to the speed of sound at a specific altitude.

Mach number A measure of the speed of an aircraft in relation to the speed of sound at a given altitude. An aircraft traveling at the speed of sound is flying at Mach 1; at half the speed of sound it is flying at Mach 0.5. The measure is named for Ernst Mach (1838–1916), an Austrian physicist.

magic window A feature of HUD and virtual cockpit displays that would use sensors to create an electronically produced image of the outside world, allowing a pilot to "see" through rain, fog, snow, and darkness. The magic window would be extremely useful in both commercial and military aircraft in all types of weather.

Maitland, Lester (1899–1990) First pilot to fly from the United States to Hawaii. With Albert Hegenberger as navigator, he successfully made the flight in a Fokker trimotor in 1927.

Mannock, Edward (1887–1918) English World War I ace who shot down 73 German planes.

Markham, Beryl (1902–1986) An Englishwoman, Beryl Markham lived most of her life in Kenya. She learned to fly in her twenties and achieved fame for her 1936 solo flight across the North Atlantic from England to Cape Breton Island, Canada. She also trained and bred racehorses and wrote a memoir, *West with the Night* (1942).

Martin In 1914 Glenn Martin built a factory for manufacturing military planes, which in 1918 produced the first American-designed bomber, the MB-1. Numerous other aircraft followed, including the B-26 Marauder of World War II. Martin began testing rocket propulsion and building missiles after World War II, and in 1961 the company merged with American-Marietta Company to form Martin Marietta, which specializes in aerospace technology.

Martin, Glenn (1886–1955) U.S. airplane designer, whose bombers and flying boats played important roles in World War II. He was a barnstormer before going into aircraft design and manufacture. His first bomber was too late for extensive use in World War I, but its success in the hands of Colonel Billy Mitchell established Martin as one of the leading military aircraft manufacturers in the United States.

McDonnell Douglas An American aerospace company formed in 1967 by the merger of the McDonnell Aircraft Corporation and Douglas Aircraft. The two companies retain their separate identities even though their products carry the corporate name. McDonnell Douglas continues McDonnell's earlier development of spacecraft and missiles, but has also designed military STOLs and attack aircraft.

McNary-Watres Act An act passed by Congress in 1930 stipulating that airlines carrying mail would be paid by the amount of space available rather than by the pounds of mail carried per mile.

Merlin engine One of the leading World War II engines, built by Rolls-Royce (and Packard, under license) and used in fighter aircraft such as the Hawker Hurricane and the P-51 Mustang.

Messerschmitt A German aircraft manufacturer founded by Willy Messerschmitt in 1923. Its Me 209 set a world speed record of 469.22 miles per hour in 1939. During World War II the company built military fighters and bombers as well as the world's first operational jet aircraft, the Me 262. Today Messerschmitt is part of Messerschmitt-Bölkow-Blohm.

Messerschmitt, Willy (1889–1978) A leading German aircraft designer before and during World War II. In 1923 Messerschmitt formed his own company, Messerschmitt AG, which by the late 1930s became the largest German aircraft manufacturer. Messerschmitt worked in Argentina after the war, but returned to Germany in the mid-1950s, where he died.

Willy Messerschmitt

metal fatigue The failure of a metal to remain intact when flexed repeatedly.

microwave landing system (MLS) An instrument landing system that operates in the microwave spectrum of radio frequencies, providing a display of horizontal and vertical lines that the pilot uses to position the aircraft when preparing for landing.

midair refueling There are two methods of midair refueling. In the probe-and-drogue method, a tanker airplane lets out a hose with a funnel, or drogue, at the end, which is attached to the probe of the receiving aircraft. In the second method, the tanker places a maneuverable boom into a receptacle on top of the receiving plane.

MiG A Soviet design bureau formed by Artem Mikoyan and Mikhail Gurevich in 1939. During World War II the team produced the MiG-1 and MiG-3 interceptors. In 1946 they built the MiG-9 fighter, one of Russia's first jet aircraft. The MiG-15 was in service with China and North Korea during the Korean War, but at least 18 other air forces have used it as well. Many more MiG models have followed.

MiG Alley The nickname for the Yalu River valley in Korea during the Korean conflict.

mission-adaptive wings One-piece wings designed to change their shape as necessary, by means of flexible composite materials and computerization, in order to maintain peak aerodynamic efficiency through all stages of a flight.

Colonel William Mitchell

Mitchell, Colonel William "Billy" (1879–1936) U.S. army officer who advocated for a separate U.S. air force and greater preparedness in military aviation. Mitchell began his military career as a private in the Spanish-American War and eventually became the outstanding U.S. combat air commander of World War I. He antagonized the military hierarchy with his outspoken calls for the establishment of an independent air force. After the loss of the dirigible *Shenandoah* in a storm, he publicly accused the war and navy departments of "incompetency, criminal negligence, and almost treasonable administration of the national defense." For this he was court-martialed, after which he resigned from the army. In 1946 the U.S. Congress authorized a special medal in his honor, which was presented to his son by the chief of staff of the newly created U.S. Air Force.

Mitsubishi A Japanese conglomerate of several dozen companies. Mitsubishi began producing aircraft just after World War I, and by World War II was building a wide variety of planes, including the well-known Zero fighter. The company reentered aviation in 1952, initially producing licensed aircraft and later its own designs.

monoplane An airplane with only one main wing.

Montgolfier, Joseph-Michel (1740–1810) **and Jacques-Etienne** (1745–1799) The Montgolfier brothers of France launched the first large-scale hot-air balloon in 1783, fueling it by burning straw and wool in a furnace below the bag. Later that year they conducted the first untethered launch with persons aboard: the 25-minute flight carried the two occupants 5.5 miles. Though famous for their balloons, the brothers made their living as paper manufacturers.

Morane-Saulnier A French aircraft manufacturer founded in 1911. Morane built high-performance monoplanes for use in World War I. At the beginning of World War II its MS. 406 was the first fighter used in large numbers by the French air force. The company closed in 1963.

National Advisory Committee for Aeronautics (NACA) A forerunner of NASA. Beginning in 1915, with the mission of solving the most pressing problems in aeronautics, NACA monitored the development of aircraft, established aviation policy, and saw to it that research needs were met. In 1958 it became NASA.

nacelle An enclosed streamlined housing on an aircraft for an engine, cargo, or crew. *See also* **pod.**

Nakajima Founded in 1917, the Nakajima Aircraft Company produced more aircraft than any other Japanese manufacturer during World War II. It built a variety of planes for the Japanese army and navy, including the Ki.-115 "Sabre" kamikaze plane and the B5N "Kate," which bombed Pearl Harbor.

National Aeronautics and Space Administration (NASA) Headquartered in Washington, D.C., NASA is mandated to research and develop flight activity within and beyond the earth's atmosphere. In 1969, NASA's Apollo spacecraft landed the first human astronaut, Neil Armstrong, on the moon. Unpiloted NASA programs such as Viking, Mariner, Voyager, and Galileo, have gathered vital statistics on other solar bodies. Communications satellites and weather satellites are just two more examples of the wide application of NASA projects.

National Air Pilots Association An organization formed in 1928 by airmail pilots who wanted to change their public image as stunt fliers to that of professional pilots.

National Air and Space Museum Established by Congress in 1946, the National Air Museum was formed to house and display historic aircraft and to educate the public about the development of aviation. (The name was amended in 1966 to the National Air and Space Museum.) The museum is part of the Smithsonian Institution.

National Championship Air Races Annual air races currently based in Reno, Nevada. First held in 1926, the races quickly caught the public eye, and manufacturers began designing more aircraft specifically for racing. Today the races also feature air shows by military aerobatic teams such as the U.S. Air Force Thunderbirds and the U.S. Navy Blue Angels.

Navy Fighter Weapons School A training program in California, founded by the U.S. Navy to train fighter pilots in the latest fighter tactics.

navigation The charting and following of a course by an aircraft or ship by use of instruments or by map reading.

navigation lights A system of lights attached to an aircraft: red on the left wingtip, green on the right wingtip, and white on the tail. These allow an observer to judge whether an aircraft is approaching or going away.

Nieuport A French manufacturer that produced record-breaking monoplanes from 1911 to 1913. Nieuport biplane fighters and bombers were used by several countries, including the United States, in World War I. Following the war the company absorbed the Astra airship company, becoming Nieuport-Delage. The company was nationalized in 1936 and became part of Société Nationale de Constructions Aéronautiques de l'Ouest.

no-man's-land The land between enemy forces in wartime.

North American Aviation (NAA) North American Aviation was formed in 1928 but was reorganized in 1934 with a manufacturing division. NAA produced fighters, trainers, and bombers used in World War II and Korea. NAA also built the world's first supersonic fighter, the F-100 Super Sabre, in 1953. NAA eventually became part of Rockwell International.

Northrop

Northrop An American aircraft manufacturer founded by John Knudsen Northrop in 1929. Following Douglas's purchase of the company in 1939, it built the P-61 Black Widow night fighter during World War II. Beginning with its 1940 production of the first flying wing aircraft, it built advanced-technology craft such as T-38 Talon supersonic trainers and remotely piloted vehicles.

John Northrop

Northrop, John (1895–1981) A co-founder of Lockheed Aircraft Company in 1927, Northrop left the following year to found a series of his own companies. He designed and built the Northrop Alpha, one of the first modern, low-wing, all-metal monoplanes. Through the 1930s and 1940s he built a variety of military planes and conceived the flying wing.

nuclear bomber An aircraft equipped to deliver nuclear weapons to a given target.

Charles Nungesser

Nungesser, Charles (1892–1927) A French World War I ace, Nungesser had transferred to aviation from the cavalry, as had the German Richthofen and the Canadian Billy Bishop. He was killed in an attempt to win the New York–Paris Orteig Prize, which Charles Lindbergh won a short time later.

oblique wing Also called a "scissor" wing, a thin wing that is pivoted to different positions on a center point to allow the aircraft to operate efficiently at both high and low speeds.

Ohain, Hans von (1911–) German aeronautical engineer whose engines powered the first successful jet aircraft, the Heinkel He-178 in August 1939. Ohain was awarded jet engine patents while still a student at Göwmlauttigen University. After the fall of Hitler's Germany, Ohain worked as a researcher for the U.S. government.

ornithopter A flying machine with flapping wings like a bird. The word comes from the Greek words for *bird* and *wing*.

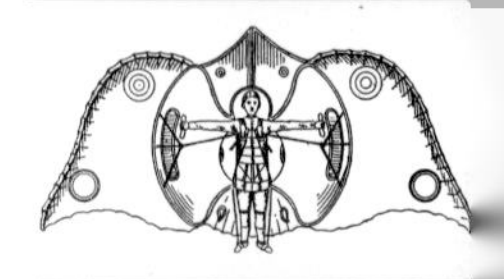
ornithopter

Orteig Prize The $25,000 prize posted after the 1919 Alcock–Whitten-Brown flight, for the first flight between New York and Paris.

Pan Am Formed in 1927 as Pan American Airways, Pan Am was the oldest international airline in the United States when it went out of business in 1991. Originally flying a mail route between Key West, Florida, and Havana, Cuba, Pan Am eventually extended its service to South America, Europe, and the Far East. In 1970 it became the first airline to use the Boeing 747 jumbo jet.

parachute A canopy of fabric panels attached to rigging lines and a harness, used for emergency aircraft evacuation, supply drops, or recreational sport. Early parachute designs date back to those of Leonardo da Vinci in the late fifteenth century.

payload Aircraft cargo that is not necessary for the operation of flight; the money-generating portion of an aircraft's load—anything from passengers to mail.

photo reconnaissance Photos taken from an aircraft that provide information about the enemy's position and strength.

Piasecki, Frank (1919–) The founder of the Piasecki Helicopter Company, manufacturer of some of the finest helicopters used during the Vietnam War.

Piper An American small-aircraft manufacturer that began as Taylor Aircraft Company in 1931, with the partnership of W. T. Piper and C. G. Taylor, producing the first model of the Cub. Following the company's reorganization as Piper Aircraft Corporation in 1937, it built a navy ambulance and a trainer during World War II. Its postwar planes included agricultural aircraft and the Cherokee series.

piston engine A common type of gasoline engine used in aircraft, particularly those manufactured before the advent of the jet engine. It uses a process of internal combustion to compress fuel in a cylinder and then ignite it with a spark plug. This explosion forces a piston downward, turning a shaft that powers the aircraft's propeller. Larger piston engines have four cycles: intake of fuel, compression, ignition, and exhaust of burned fuel. Smaller engines, such as those in lawnmowers or outboard motors, usually accomplish this in two cycles.

pod The structure encasing an externally mounted engine; or a compartment attached to the fuselage holding fuel, radar equipment, or weapons.

Post, Wiley (1899–1935) One of the most colorful figures of early U.S. aviation. He made the first solo flight around the world, designed and successfully used the first pressurized flight suit, and proved the value of navigational instruments, including the automatic pilot. Two years after his around-the-world solo flight, he and his passenger, the humorist Will Rogers, were killed in a plane crash in Alaska.

Pratt & Whitney An American builder of aircraft engines. Today part of United Technologies Corporation, Pratt & Whitney was founded in 1925 as a specialist in air-cooled radial engines. Early successes were the Wasp (1926) and the Hornet (1927). More recent engines include the JT3C two-spool turbojet, the JT3D turbofan, and the JT9D, installed in the Boeing 747 and DC-10.

precision bombing A bombing strategy that targets particular structures, such as factories or power plants.

pressurized aircraft Aircraft in which compressed air is circulated throughout the cabin during flight. Pressurization protects the crew and passengers from the dangerous effects (ranging from ear discomfort to unconsciousness) of the low pressure and oxygen deficiency of high altitudes.

pressurized suit A sealed suit designed to maintain normal pressure on a pilot's body to compensate for the reduced atmospheric pressure of high-altitude flight. Wiley Post successfully used the first pressurized suit in 1934.

propeller A mechanical device having one or more blades rotating about a central shaft. In the same way that a wing in motion lifts an airplane, the propeller blades in motion pull (or push) it forward.

Pulitzer Race With the Schneider Cup, one of the two most important speed trophies. Both were won by American pilots in 1925.

pusher An airplane with the propeller

mounted behind the wings, to push the airplane forward. Curtiss produced several pusher engine planes for the U.S. Navy from 1910 to 1918.

Harriet Quimby

Quimby, Harriet (1884–1912) A Boston woman who was the first U.S. licensed woman pilot in 1911. She was also the first woman to fly the English Channel.

R

radar An acronym for "radio direction and range," radar is a system that uses shortwave radio transmissions to determine the distance of clouds, physical features on land, or other aircraft. Radar is crucial to air traffic control systems, but it is also used in combat to locate targets and to gain advance warning of enemy attack. The basic idea of radar dates to the late 1880s in Germany, but it was not set in operation until the start of World War II in 1939.

radarscope The screen that visually displays reflected radio beams picked up by a radar receiver. Types of displays include a Plan Position Indicator with a rotating scanner beam (used in navigational radar), a digital display that transmits a variety of aircraft information (used in aircraft control systems), or an actual map that locates targets and enemy aircraft (used in combat).

radio compass A direction finder used in navigation.

ramjet engine An engine that eliminates the compressor and turbine from the turbojet engine, converting the speed of incoming air into pressure using a "ram" effect. Whereas a turbojet engine can attain a maximum speed of Mach 3 to 3.5, a ramjet can reach Mach 5 to 6.

reconnaissance An exploratory military survey of enemy territory and of enemy installations, movements, activities, and strength.

remotely piloted vehicles (RPVs) Aircraft that can be controlled in flight without a pilot on board. RPVs may be used as spy planes to gather information on ground targets, enemy communication, and radar systems, and they may even drop bombs or missiles. They can also relay communications, detect mines, and gather weather data.

Republic Founded as the Seversky Aircraft Corporation in 1931 and renamed Republic Aviation Corporation in 1939, this company produced a variety of military aircraft, including the Thunderbolt fighter (1941) and Thunderchief fighter-bomber (1955). The company was purchased in 1965 by Fairchild-Hiller and became Fairchild-Republic, which today is part of Fairchild Industries.

Richthofen, Manfred von (1892–1918) Germany's top aviator and leading ace in World War I. He was commander of Jagdstaffel 11, known as "Richthofen's Flying Circus" because of its fancifully decorated scarlet planes, and he became known as the "Red Baron." He was credited with shooting down 80 enemy aircraft. Richthofen decorated his room with "souvenirs" from the airplanes he shot down. After his death he was replaced as commander of the fighter group by Hermann Göring.

Rickenbacker, Eddie (1890–1973) The most celebrated U.S. ace of World War I. He had been an auto racer and became a fighter pilot when the United States joined the Allies in 1917. Although the United States entered the war rather late, he accumulated 26 air victories. After the war he became involved with automobile manufacture and with commercial airlines, eventually becoming president, general manager, and director of Eastern Air Lines.

Rockwell In 1967 Rockwell merged with North American Aviation to form North American Rockwell. Six years later Rockwell International was established. The company produced the B-1 bomber, a long-range, supersonic aircraft. It also produces civil planes such as Commander twins and Thrush Commander agricultural aircraft.

Rodgers, Calbraith (1879–1912) First U.S. aviator to make a transcontinental flight. When the Hearst newspapers offered a $50,000 prize to the first pilot to make a cross-country flight in less than 30 days, Rodgers, who had trained at the Wrights' flying school, decided to make the attempt in a Wright biplane he named the *Vin Fiz*. In his 1911 flight, he succeeded in flying across the country but did not win the prize because the trip took 49 days.

Rolls-Royce This European automobile maker, founded in 1906, began the manufacture of renowned aircraft engines in 1915. Notable among them are the Merlin (1933); the Welland (1943), the company's first jet engine; and the Avon (1959), a turbojet. Rolls-Royce absorbed Bristol-Siddeley Engines in the late 1960s, but since its own bankruptcy in 1971, its engine divisions have been owned and operated by the British government.

rotor A system of horizontal blades that provide the force to support an aircraft in flight. A tail rotor also provides directional control.

Royal Flying Corps Predecessor of the Royal Air Force, formed in England in 1912 with naval and military wings and a central flying school. In 1914 the naval wing split off and became the Royal Naval Air Service, while the military wing kept the title Royal Flying Corps. In 1918 the two organizations were absorbed into the Royal Air Force, independent of the navy and the army.

rudder A movable airfoil, usually attached at the rear of an airplane, that causes the plane to turn left or right horizontally.

runway A level strip of usually paved ground used for the takeoff and landing of aircraft.

Rutan, Burt (1943–) An aircraft designer whose work has influenced both commercial and homebuilt designs. Rutan has shown that canards can be used successfully on modern aircraft. He has also been instrumental in constructing planes using fiberglass on foam because of its strength, low cost, and smooth surfaces that reduce drag.

Rutan, Richard G. (Dick) (1938–) Retired U.S. Air Force colonel and older brother of Burt Rutan. Cofounder, with Jeana Yeager, of Voyager Aircraft, Inc., a company formed to build a graphite-composite aircraft designed by Burt Rutan. In late 1986, Dick Rutan and Yeager, copiloting *Voyager*, completed the first nonstop, unfueled flight around the world, in nine days.

Ryan: *Spirit of St. Louis*

Ryan Originally called Ryan Airlines, Ryan Aircraft was established by Tubal Claude Ryan in 1922. Its early products included the *Spirit of St. Louis*, built for Charles Lindbergh. The company went out of business during the Depression, but Ryan founded Ryan Aeronautical in 1933 to produce trainers. The company also built the FR-1 Fireball fighter (1944) and several V/STOL aircraft before joining Teledyne in 1969. Today Teledyne Ryan produces remotely piloted vehicles.

Santos-Dumont airship

Santos-Dumont, Alberto (1873–1932) A Brazilian-born sportsman and scientist. An important figure in the early popularization of aviation, he flew for the joy of it. He built and flew his first small airship in 1898 and continued experiments in lighter-than-air craft and, later, heavier-than-air craft well into the first decade of the twentieth century. In 1904, when he turned to heavier-than-air flight, he had become a successful builder of nonrigid airships.

Saulnier, Raymond French aviator and inventor who developed deflector plates for an airplane's propeller, which prevented a machine gun's fire from damaging the propeller. This invention was superseded by the interrupter gear of the Fokker Eindecker.

Savoia-Marchetti An Italian aircraft manufacturer that came to prominence in the early 1920s with its production of flying boats and then expanded into trimotor planes in the 1930s. Also known as SIAI-Marchetti, the company resumed production after World War II, building private aircraft as well as military trainers and utility planes.

Schneider Trophy A trophy for top speed for seaplanes in international competition, established in 1913. Britain permanently acquired the trophy after its 1931 win, with a speed of 340 miles per hour, having won it three times in succession.

scramjet engine Short for "supersonic combustion ramjet," the scramjet engine mixes incoming air with fuel and ignites the mixture while the air is still moving at supersonic speed. (Turbojet and ramjet engines must slow down the speed of incoming air.) Scramjets have been shown to attain speeds of Mach 8.

seaplane An aircraft that can land, float, and take off on water. Those with boatlike hulls are called flying boats, and those with separate pontoons are called floatplanes. Seaplanes were first built and flown in the United States by Glenn Curtiss in 1911 and 1912, and by the late 1920s the largest and fastest planes in the world were seaplanes. Their importance diminished after World War II because of the increased range of land-based planes, the construction of land bases, and the development of aircraft carriers.

self-repairing control system Currently being developed by the U.S. Air Force, this system would use fly-by-wire technology to maintain flight control if an aircraft's primary control surfaces were severely damaged or blown off in combat. In such a situation, the system would reconfigure itself so that the plane could function adequately with the remaining control surfaces (such as rudders and ailerons).

sesquiplane A biplane with a lower wing of less than half the area of the other.

Sikorsky Established as a U.S. company by Russian-born Igor Sikorsky in 1923, the Sikorsky Aero-Engineering Corporation initially produced flying boats. In the late 1930s it began building very successful helicopters, which became widely

Igor Sikorsky

(1889–1972) Russian-born U.S. pioneer in aircraft design, best known for his successful development of the helicopter.

He began work on helicopters in Russia in 1909 but was unsuccessful because the engines available at the time were too heavy. He switched to fixed-wing aircraft and produced the LeGrand, an all-enclosed, 4-engine aircraft, which he flew successfully in 1913. He produced successful bombers for Russia in World War II. Sikorsky became a U.S. citizen in 1928 and set up an aircraft factory with the financial backing of Sergey Rachmaninoff, the famed Russian concert pianist and composer.

Sikorsky returned to his work on the helicopter in the 1930s. His first U.S. helicopter flew successfully on its first test flight in 1939.

used by the military in World War II as well as in Vietnam.

single-seater An airplane with a single seat.

Skunk Works The research and development subsidiary of Lockheed Corporation, located in the Mojave Desert. Founded in 1943 by Clarence "Kelly" Johnson, Skunk Works' biggest successes to date have been the XP-80 Shooting Star (the first operational U.S. fighter jet), the famous U-2 and SR-71 spy planes, and the F-117A stealth fighter used in the Persian Gulf War.

slipstream The flow of air generated by an aircraft's propellers. In single-engine aircraft, this fast-moving column of air is pushed back down the fuselage, causing the aircraft to swing, or yaw.

smart skin The sensor-equipped outer surface of future fighter aircraft. Fighters will be extensively composed of composites that will allow the "skin" of the aircraft to collect important information.

Smith, Lowell (1892–1945) The flight commander of the around-the-world flight of the Douglas World Cruisers in 1924. He also pioneered the procedure of midair refueling.

sonic boom A sound similar to an explosion, caused when the shock wave that forms at the nose of an aircraft traveling at supersonic speed reaches ground level.

Sopwith A British aviation company founded at a converted ice rink by T. O. M. Sopwith in 1912 and managed by Sopwith, designer Fred Sigrist, and test pilot Harry Hawker. Sopwith produced a number of military aircraft in World War I, including the well-known Camel. The company closed in 1920.

Sopwith Camel

Sopwith, T. O. M. (1888–1989) A British aircraft designer whose Sopwith Aviation Company (formed in 1912) produced the World War I Sopwith Camel. Sopwith was also a prize-winning pilot, capturing the first aerial Derby in a Blériot plane in 1912. After his company closed in 1920, Sopwith became chairman of the Hawker Siddeley Group from 1935 to 1963.

sortie A single attack or mission by an airplane.

sound barrier The sudden, extreme drag an aircraft encounters when approaching supersonic speed. When an aircraft travels at less than the speed of sound, pressure waves in front of the aircraft push air out of the way. When the aircraft reaches the speed of sound, however, the aircraft is traveling as fast as the pressure waves, so it encounters a barrier, or "wall," of air that produces turbulence and increased aerodynamic drag.

SPAD A French aircraft manufacturer, formed when a group of businessmen, led by Louis Blériot, took over Depperdussin in 1914. It produced military reconnaissance planes and fighters such as the SPAD XIII, which was flown by the British, Americans, Italians, and Belgians, as well as the French, during World War I.

spat

spat Wheel coverings on the landing gear of certain planes.

stall A loss of lift. When an airplane's angle of attack is so steep that the airflow over the upper surface of the wing breaks, lift is lost and drag increases.

stealth aircraft Aircraft that are capable of penetrating enemy defenses without being detected. "Low observables," or features that decrease the chances of being detected on enemy sensors, make an aircraft more "stealthy." Low observables include a "blended" aircraft body with no sharp corners that would easily reflect radar signals; a fuselage with no gaps, cracks, or corners; and weapons carried internally, rather than externally. The U.S. military is extremely secretive about the details of the stealth aircraft currently in development or in use, because revealing details of their characteristics (even by flying the planes in public) could reduce their effectiveness against a potential enemy.

Katherine Stinson

Stinson, Katherine (1891–1977) Well-known exhibition flier before World War I. In 1913 she became the first woman to fly the mail. She toured extensively in England, China, and Japan, giving flight exhibitions.

strafing Raking with fire at close range, especially machine-gun fire from low-flying airplanes.

Stringfellow, John (1799–1883) Stringfellow, with his colleague, William Samuel Hensen, contributed to the development of modern airplanes, primarily through their models and drawings. Stringfellow's 1868 powered model triplane, exhibited at the British Aeronautical Society, was widely publicized and had great impact. It involved separate systems for lift, propulsion, and control.

strut A brace that supports a load, for instance, the structure supporting a wing.

Stuka dive bombers A dive bomber developed beginning in 1934 and first used in the Spanish civil war at the end of 1937. The Stuka earned a deadly reputation in blitzkrieg attacks by the Luftwaffe in World War II, particularly against Poland and France in 1939 and 1940. The last Stukas were produced in 1944.

Sud-Aviation A French aircraft manufacturer

formed in 1957 through the merger of SNCASE and SNCASO. Sud-Aviation originated the Caravelle jet airliner and built numerous helicopters and light aircraft. The company merged with Nord-Aviation and SEREB in 1970 to form Aérospatiale, one of the largest European aerospace companies.

Sukhoi A Soviet design bureau established in 1936 by Pavel Osipovich Sukhoi. In 1933 Sukhoi had designed the I-14, the first Soviet fighter to have an enclosed cockpit and retractable landing gear, as well as record-breaking long-distance aircraft, and jet bombers and fighters.

superalloy A blend of two or more metals to make a new, stronger metal that will maintain its strength and other desirable properties under high temperatures.

supercharger A device that increases the volume of air drawn in by an internal-combustion engine to compensate for the lower density of the air in high-altitude flight.

supermaneuverability The capability of a fighter to carry out maneuvers at angles of attack beyond the point at which stall occurs. With vectored-thrust engines and other modified features, an aircraft could slow down so suddenly it would seem to have stopped flying, or it could move sideways without changing its nose direction. *See also* **vectored thrust.**

Supermarine A British manufacturer established just before World War I to produce marine aircraft and amphibians. During the 1920s it also developed seaplanes for the Schneider Trophy races. The Spitfire of World War II is Supermarine's most famous product. After World War II Supermarine built the Swift and other fighters, becoming part of the British Aircraft Corporation in the early 1960s, along with its parent company, Vickers.

supersonic flight Flight beyond Mach 1, or the speed of sound, at a given altitude. At supersonic speeds, an aircraft encounters less turbulence and aerodynamic drag than at the sonic barrier, because shock waves produced by the speed are stabilized. Charles Yeager of the U.S. Air Force is given credit for first attaining supersonic speed when he reached Mach 1.06 at 45,000 feet in 1947. *See also* **sound barrier.**

supersonic transport (SST) An airliner designed to fly at supersonic speeds. The best-known SST is the Concorde, built by the British and the French, which flies at about 1,400 miles per hour (a little above Mach 2). SSTs of greater Mach speeds have been proposed, but environmental concerns of excessive noise and further destruction of the earth's ozone layer have delayed further planning.

swept-back wings

swept-back wings Wings of an aircraft that have a backward slant from the point where the wing joins the body of the aircraft to the wing tip.

swept-forward wings

swept-forward wings Wings that slant forward toward the nose of an aircraft.

T

tachometer An instrument that indicates the rate of a rotating shaft by its revolutions per minute (rpm).

taxiway An airport roadway used by aircraft traveling between a terminal or hangar and a runway.

terminal A freight or passenger station that serves as an aircraft servicing facility as well as an end or junction point for air travel.

Louise Thaden

Thaden, Louise (1905–1979) Winner of the Bendix Trophy in 1936, Louise Thaden set a number of records during her career as an aviator, including women's altitude, speed, and endurance records.

three-axis control Control of the rotation of an aircraft about all three axes, giving stability and navigational ability. Rotation about the horizontal axis from nose to tail is called *roll.* It is controlled by wing warping or by ailerons and is required in order to bank the aircraft on turns. Rotation about the vertical axis is called *yaw.* It is controlled by the rudder and makes turning possible. Rotation about the horizontal axis perpendicular to the other two axes is called *pitch.* It is controlled by the elevator and makes ascent and descent possible.

thrust The forward moving force produced by a propeller or a jet or rocket engine as exhaust gases are ejected through the exhaust nozzle.

tilt-rotor V/STOL A type of V/STOL that can take off like a helicopter and fly like an airplane. With an airframe constructed almost entirely of composites and a fly-by-wire control system, the V-22 Osprey (currently the best-known model of this aircraft) can reach speeds above 300 miles an hour.

titanium A chemical metal element whose properties of light weight, high strength, and low corrosion make it ideal for use (usually in alloy form) in parts for high-speed aircraft.

Top Gun school *See* **Navy Fighter Weapons School.**

torque The tendency of an aircraft to rotate in the direction opposite to that of the propeller, especially on takeoff.

torsion The twisting or wrenching of a body by forces that turn one end or part while the other is held fast or turned in the opposite direction.

trailing edge The rearmost edge of an airfoil.

transport plane An aircraft designed to carry people or cargo.

Travel Air An American aircraft manufacturing company founded in 1924 by Walter Beech. Travel Air built airplanes for sport and business purposes, and several models won racing prizes, including the Dole Prize in 1927 and the Free-for-All (later the Thompson Trophy Race) at the 1929 National Air Races. Travel Air merged with Curtiss-Wright in 1930.

Trenchard, Sir Hugh (1873–1956) An English flier who rose through the ranks of the air corps to become chief of staff (1919), and later the first marshal (1927), of the Royal Air Force. His policy of launching persistent air attacks in combat became a standard practice of the RAF.

trim tabs Small surfaces that mechanically or electronically manipulate the rudder, elevator, and ailerons to help stabilize the plane. Trim tabs free the pilot from constantly adjusting the controls.

triplane An aircraft with three wings of uniform size, stacked vertically one above the other. The design was often used in World War I fighter planes because it offered maneuverability and a short takeoff run.

Trippe, Juan (1899–1981) A U.S. World War I pilot who in 1925 helped to form Colonial Air Transport, which flew the first contract airmail service between New York City and Boston. In 1927 Colonial Air merged with two other small airlines to form Pan American Airways, for which Trippe served as president until his retirement in 1968.

Tupolev A Soviet design group, led by Andrey Nikolayevich Tupolev, that was prominent from the 1920s through the 1960s. Tupolev produced more than one hundred types of passenger and military aircraft, including many of Russia's heavy bombers.

turbofan engine An engine similar to a turbojet engine, with the exception that a portion of the air drawn into the engine bypasses the compression process, passing instead through a fan. This air may then be ejected as a "cold" jet or may be mixed with the turbine exhaust, becoming a "hot" jet.

turbojet engine An engine that operates by burning a mixture of compressed air and fuel, and then forcing the hot air through a turbine (which drives the compressor) and through an exhaust nozzle to produce a "jet," or stream of gas; the reaction to it provides thrust for an aircraft.

turboprop engine An engine similar to a turbojet engine, with the exception that the pressure of the air and gas mixture passing through the turbine, or turbines, produces enough power to drive both the compressor and a propeller. Aircraft with turboprop engines thus obtain thrust both from the jet and from the propeller.

turbosupercharger A turbine pump or compressor, driven by the engine exhaust gases, that increases the density of the air or the air-and-gas mixture in an aircraft engine, so that the resulting exhaust is powerful enough to drive both the turbine and a rotor for additional power.

Roscoe Turner

Turner, Roscoe (1895–1970) A U.S. racing pilot who helped aviation become popular in the 1920s and 1930s. Turner established one of the first U.S. airlines in 1929, flying between Reno, Nevada, and Los Angeles, California. Through the 1930s he flew record-setting races in airplanes that he helped to design and build. Later he founded Roscoe Turner Aeronautical Corporation and Turner Aviation Corporation.

TWA An airline formed as Transcontinental and Western Air in 1930, although in 1950 its name was changed to Trans World Airlines. Initially offering coast-to-coast service, TWA expanded to provide international service to countries throughout the world.

U

ultralight A very light, low-speed airplane, powered by a small engine. Ultralights may be launched by foot or may have light landing gear. The popularity and complexity of ultralights and other experimental aircraft continue to grow.

United Aircraft and Transport Corporation A company formed in 1928 by the merger of numerous companies, including Pratt & Whitney, Boeing, and several airlines that later became United Airlines. In 1934 the new merger was broken up by Congress into three companies: United Aircraft Corporation (later to become United Technologies Corporation), Boeing, and United Airlines.

U.S. Strategic Air Command (SAC) One of the three major combat commands of the U.S. Army Air Forces, formed in 1946. Although SAC was armed with nuclear weapons, and by the late 1950s was capable of attacking a target anywhere in the world, it served primarily as a deterrent force.

WASP insignia

U.S. Women's Airforce Service Pilots (WASP) An organization of women pilots in the military, formed in 1943. The WASPs' duties included ferrying aircraft, towing targets in fighter pilots' training, and test-flying aircraft, among them the famous Flying Fortress. The program was deactivated in 1944, but the WASPs were granted a belated veterans' status in 1977.

V

vectored thrust The thrust provided from an engine equipped with exhaust nozzles that can be swiveled downward, to the rear, or even slightly forward to give an aircraft thrust in a variety of directions, resulting in one form of supermaneuverability. Vectored thrust is an important feature of the X-31.

vertical flight The capability of an aircraft to take off and land vertically. It is also referred to as V/STOL for Vertical/Short Takeoff and Landing, or VTOL, for vertical Takeoff and Landing—in which the

U.S. Air Force

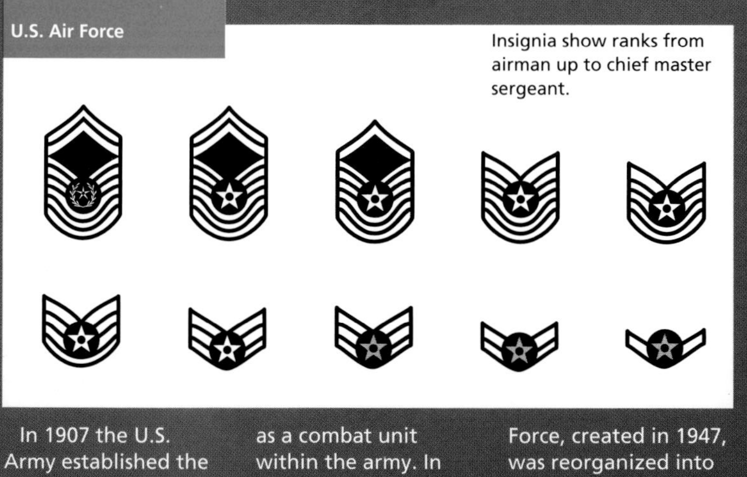

Insignia show ranks from airman up to chief master sergeant.

In 1907 the U.S. Army established the Aeronautical Division to oversee military ballooning and aircraft. By the end of World War I, the Army Air Service (later called the Air Corps) had been established as a combat unit within the army. In 1941, with the entry of the United States into World War II, the Air Corps became the Army Air Force, which concentrated on strategic bombing. An independent U.S. Air Force, created in 1947, was reorganized into the Department of the Air Force within the Department of Defense in 1949. The headquarters of the Air Force are at the Pentagon, outside Washington, D.C.

craft needs no runway space to take off and land.

Vickers A British manufacturer of armaments, founded in 1911. In 1938 Vickers merged with Supermarine to form Vickers-Armstrong, which produced the Wellington bomber between 1938 and 1945, as well as the Viscount, the world's first turboprop airliner. In the early 1960s the company was absorbed into the British Aircraft Corporation.

Vin Fiz The name of the Wright biplane flown across the continent in 1911 by Calbraith Rodgers.

virtual cockpit A head-up display (HUD) being developed for military aircraft that would display crucial information about an aircraft's systems through the pilot's helmet. In combat, the virtual cockpit would give the pilot instant access to this information as well as a 360-degree view of targets and enemy aircraft.

Voisin The name of aircraft built by French brothers Gabriel (1880–1973) and Charles (1882–1912) Voisin. The world's first successful amphibian plane (1912) and the first airplane with an all-metal frame (1913) were Voisins. Gabriel Voisin managed the construction of more than ten thousand warplanes for Allied forces in World War I.

Voyager An American experimental aircraft, designed by Burt Rutan and piloted by Dick Rutan and Jeana Yeager, that in 1986 became the first airplane to fly around the world without stops or refueling. Before its flight, the fuselage, wings, and other frame areas were entirely filled with fuel that weighed four times as much as the plane itself. Only a few gallons of fuel remained unused when the plane landed.

V/STOL aircraft (Vertical/Short Takeoff and Landing) Aircraft that can take off and land vertically or within a short runway space. Although helicopters were the first V/STOL aircraft, their limits of range and speed have led designers to develop other, more flexible V/STOLs. In the future, V/STOL fighters will be able to fly slowly and hover, and will operate from severely damaged runways close to the battlefield. Tilt-rotor V/STOLs will operate like fast helicopters, carrying out a variety of missions.

VTOL aircraft (Vertical Takeoff and Landing) Aircraft that can take off and land without runway space. Sometimes VTOL craft is referred to under the broader term **V/STOL.**

W

WASP *See* **U.S. Women's Airforce Service Pilots.**

Whirlwind engine An air-cooled, radial engine produced by the Wright company beginning in 1925. This reliable and influential engine was used in light transport and private airplanes, including the one Charles Lindbergh flew solo across the North Atlantic.

Whittle, Sir Frank (1907–) British aeronautical engineer and one of the pioneers of the jet engine. His engines were used in the first British jet aircraft, the Gloster E-28/39, in 1941. After receiving a patent in 1930, Whittle's jet engine design languished. But Whittle renewed the patent and formed Power Jets Ltd. His design became a reality in 1937 with the testing of the first operational jet engine.

windscreen A screen that protects against the wind; a windshield.

wind shear A sudden change in the direction of the wind, especially the vertical shifts that aircraft can encounter near a runway.

wind tunnel A tunnel-like passage through which air is blown at a known velocity to determine the effects of wind

wind tunnel

pressure on an object such as an airplane model or part placed in the passage.

winglets Small upward projections located at the tips of an aircraft's main wings to improve the efficiency of the wings.

wing root The point where the wing joins the fuselage or the opposite wing.

wingtip-vortex turbines Turbines attached to the tips of an aircraft's main wings that reduce drag during flight and that use the wingtip vortices (the swirling air at the wingtips) as an energy source. These turbines can be connected to generators that provide electricity to be used aboard the aircraft.

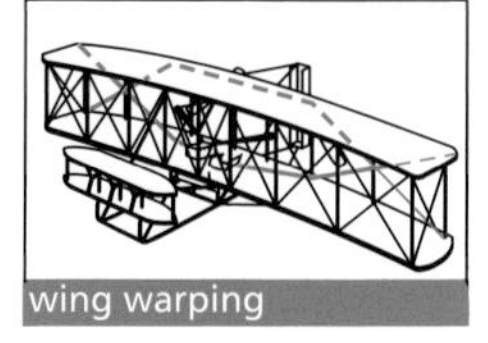
wing warping

wing warping A system using control wires by which a wing could be twisted, or "warped," to give control over banking or roll. This system, developed by the Wright brothers, was eventually replaced by ailerons.

Wrights' bicycle shop

Wright, Orville (1871–1948) **and Wilbur** (1867–1912) U.S. inventors and researchers, who first achieved piloted powered flight. In 1899, after becoming interested in bird flight and reading of Langley's experiments, the Wright brothers, who ran a printing business and a bicycle shop, decided to make a serious study of aviation. They studied, experimented extensively, and kept meticulous records. They believed it was essential to master glider flight before attempting powered flight. Their great early achievement was the control they obtained through the combination of a horizontal elevator, a vertical rudder, and wing warping. They designed and built the engine used on the first successful powered flight in 1903.

XR series A series of notable helicopters developed by Igor Sikorsky, a Russian-born engineer, for use in World War II. In just three years and four months, Sikorsky's division of the United Aircraft Corporation produced 600 helicopters in the series.

X-wing A type of V/STOL equipped with a rotor, for helicopter-like flight, that can be locked in a stationary position to serve as a fixed wing for fast, forward flight.

Yakovlev A Soviet design bureau, led by Alexander Sergeevich Yakovlev, that became prominent during World War II. Yakovlev established a lightplane factory in 1935. He built sportsplanes and trainers, but it was his fighters, particularly the Yak-9, for which he was best known. He also designed the Yak-24 "flying wagon" helicopter.

Yeager, Charles "Chuck" (1923–) An officer in the U.S. Air Force best known for being the first pilot to exceed the speed of sound, which he did in 1947 while testing the Bell X-1 aircraft. Yeager served as a test pilot from 1947 to 1954. He retired as a brigadier general in 1975 and published his autobiography, *Yeager,* in 1985.

Yeager, Jeana (1952–) Pilot and cofounder, with Dick Rutan, of Voyager Aircraft, Inc. Yeager and Dick Rutan had set aviation records flying Burt Rutan's designs before doubling the existing nonstop flight distance record on *Voyager* in 1986.

Z

Zeppelin A rigid, lighter-than-air craft of the type developed by Count Ferdinand von Zeppelin of Germany. Typical was the L-3, which served the German navy at the outbreak of World War I. It was 490 feet long, utilized three 210-horsepower engines, could cruise at 47 miles per hour, and could fly as high as 6,000 feet. *See also* **dirigible.**

Count Zeppelin

Zeppelin, Count Ferdinand von (1838–1917) German airship designer and manufacturer. Count Zeppelin visited the United States and observed the use of Army balloons during the Civil War. He made his first ascent in a balloon at St. Paul, Minnesota, in 1863 and launched his first airship in Germany in 1900. In 1906 he made a 24-hour flight, which resulted in his receiving a commission to build an entire fleet of airships. More than one hundred Zeppelins were used by the Germans in World War I. The era of rigid airships ended with the *Hindenburg* explosion and fire at the United States Naval Air Station at Lakehurst, New Jersey, in 1937.

A

Aces, 46–47
Advanced tactical fighter (ATF), 204
Aerial Experiment Association (AEA), 34
Aerial photography, 43–44
Aero Club of America, 32, 35
Aerodrome, 22–23
Aerodynamics: *See* Aerodynamic Theory
Aeromarine West Indies Airways, 96
Aeronca Champion, 104
Ailerons, 37, 55
Airacomet. *See* Bell aircraft: P-59 Airacomet
Airborne warning and control systems (AWACS), 148
Airbus Industries aircraft: A300, 168 Airbus 319, 145; Airbus 330, 207; Airbus 340, 207
Air Commerce Act, 98
Aircraft carriers, 121
Aircraft Transport and Travel, 94–95
Air France, 95
Airmail (U.S.), 90–94, 96–100; Air Mail Act (Kelly Bill), 97; McNary-Watres Act, 99
Airports, 98, 166
Airscrew, 173
Airships. *See* Dirigibles
Air Transport Auxiliary (ATA), 133
Airways, lighting, 91–92, 94
Albatros aircraft, 42, 47; D I, 52; D.Va, 54–55
Alcock, John, 72
Alpha. See Northrop Alpha
American Airlines, 99–100
Aquila, 205
Archdeacon-Deutsch prize, 38
Arnold, Henry, 127
Around-the-world flights, 83–87, 184–185
Atlantic flights. *See* Transatlantic flights
Atomic bomb, 138–139
Auditorium Hotel Trophy, 73
Autogyro, 174
AV-8B Harrier. *See* British Aerospace Harrier
Aviation Corporation (AVCO), 99
Avionics, 198, 200–201; fly-by-wire, 170, 189
Avro aircraft, 42; 504, 66; Lancaster, 134

B

B-17 Flying Fortress. *See* Boeing aircraft
B-24 Liberator. *See* Consolidated Aircraft Corporation
B-25 Mitchell bomber. *See* North American aircraft
B-29 Superfortress. *See* Boeing aircraft
B-47. *See* Boeing aircraft
B-52. *See* Boeing aircraft
B-58 Hustler. *See* Convair aircraft
Balchen, Bernt, 71
Balloons, 18, 44. *See also* Dirigibles
Barnstormers, 65–66
Beachey, Lincoln, 64, 65
Beaverbrook, Lord, 116
Beech aircraft: Beechcraft Bonanza, 104; Beechcraft Starship 2000A, 194; Staggerwing, 69; Travel Air, 99
Beech, Walter, 99
B. E. I. biplane. *See* de Havilland aircraft
Bell aircraft: HueyCobra, 177; JetRanger, 177; P-59 Airacomet, 144; X-1, 141, 151
Bell, Alexander Graham, 34, 180
Bell-Boeing V-22 Osprey, 206
Bendix Trophy, 67, 69, 73
Bennett, Floyd, 71
Berlin airlift, 155–156
Bishop, Billy, 47
Blériot, Louis, 39 Blériot XI, 38, 39, 42
Blitzkrieg, 108
BMW engine, 56
Boeing aircraft: B-17G Flying Fortress, 122, 127, 128, 130–131, 134; B-29 Superfortress, 137, 138; B-47, 153–155; B-52, 154–155; Boeing 247, 101; Boeing 314, 68, 101; Boeing 707, 158, 162–163; Boeing 727, 165; Boeing 747, 145, 150–151, 165, 167–168, 208; Boeing 777, 206–207; CH-46 Sea Knight, 177; Monomail, 101
Boeing, William E., 99
Boelcke, Oswald, 50, 51
Bolori, Denis, 16
Bombardment, aerial: area, 127, 134; and dirigibles, 57–58; radar and, 132; World War I, 57–61; World War II, 126–129, 134
Bomber planes: jet, 153–155; World War I, 57–61; World War II, 128–129
Bonanza. *See* Beech aircraft
Bong, Richard, 119
Boulton-Paul Defiant, 115
Braniff, 100
Bristol Fighter (Brisfit), 52
British Aerospace Harrier, 190, 201
British Expeditionary Force (BEF), 43
British Overseas Airways Corporation (BOAC), 157
Brown, Walter Folger, 99–100
Bullet-proof windshield, 116
Byrd, Richard, 71, 78

C

C-2. *See* Fokker aircraft
C-5A Galaxy. *See* Lockheed aircraft
C-47. *See* Douglas aircraft
C-53. *See* Douglas aircraft
C.202 Folgore. *See* Macchi aircraft
C-54. *See* Douglas aircraft
Camel. *See* Sopwith aircraft
Canards, 194–195, 201
Caproni aircraft: Ca-36, 59; Ca 60 Transaereo, 180–181
Caproni, Giovanni, 180
Cargo: commercial service, 104, 205; military planes, 155–156
Cars, flying, 179, 181
Cayley, George, 19, 173
Cessna, Clyde, 99
Cessna aircraft: Cessna 150, 104; Citation, 165, 196
CH-46 Sea Knight. *See* Boeing aircraft
Chain Home system, 117
Challenger, 208
Chanute, Octave, 21, 25, 27
Charles, J. A. C., 18
Chennault, Claire, 125
Churchill, Winston, 119, 128
Citation. *See* Cessna aircraft
Civil Aeronautics Act of 1938, 100
Cochran, Jacqueline, 69, 133
Coffyn, Frank, 64
Combined-cycle engines, 196–197
Comet. *See* de Havilland aircraft
Commercial aviation: airmail, 90–94 cargo service, 104; passenger service, 94–103
Composite materials, 182
Computer-aided design (CAD), 191
Computer-aided three-dimensional interactive application (CA-TIA), 207
Concorde, 169–171
C-130. *See* Lockheed aircraft
C-141A Starlifter, 158. *See* Lockheed aircraft

Consolidated Aircraft Corporation: B-24 Liberator, 127, 128
Constellation. *See* Lockheed aircraft
Control, in flight, 19, 20–21, 25–26, 36–37
Convair aircraft: B-58 Hustler, 154–155; F-102 Delta Dagger, 150; F-106 Delta Dart, 150; ConvAirCar, 179
Cornu, Paul, 174
Corsair. *See* Vought F4U Corsair
Crop dusting, 104
Curtiss aircraft: Hawk 75, 109; JN-4D (Jenny), 60, 65, 90; June Bug, 34; NC, 68; P-40 Warhawk, 122, 125, 135; Rheims Racer, 34; R2C-1, 67; R3C-2, 68, 70; R-6, 67; Type 75, 96
Curtiss, Glenn, 32, 34–36; exhibition team, 64
Cygnet II, 180

D

Daedalus, 16
Daimler-Benz engines, 110
d'Arlandes, Marquis, 18
da Vinci, Leonardo. *See* Leonardo da Vinci
DC-2. *See* Douglas aircraft
DC-3. *See* Douglas aircraft
DC-8. *See* McDonnell Douglas aircraft
DC-10. *See* McDonnell Douglas aircraft
D-Day, 134
de Bothezat, George, 174
de Havilland aircraft: B. E. I biplane, 38; Comet, 159–161; D.H.2, 50; D.H.4, 58, 90, 95; F.E.2b, 50, 58
de Havilland, Geoffrey, 38, 50, 157
de la Cierva, Juan, 174
Delta Airlines, 100
Delta Dagger. *See* Convair aircraft
Delta Dart. *See* Convair aircraft
Deutsche Luft Hansa. *See* Lufthansa
Dewoitine 520, 109
D.H.2. *See* de Havilland aircraft
D.H.4. *See* de Havilland aircraft
D I. *See* Albatros aircraft D I
di Cosimo, Piero, 16
Digital flight-control system (DFCS), 204
Dirigibles, 44–45, 57–58, 105, 201
Distinguished Flying Cross, 122
Doolittle, James, 68, 70, 123
Dornier bombers, 119
Douglas aircraft: C-47, 101–103; C-53, 102–103; C-54, 156–157; DC-2, 101; DC-3, 101–103; M-2, 90; TBD Devastator, 125; World Cruiser, 83–84. *See also* McDonnell Douglas aircraft
Dowding, Sir Hugh, 112, 116
Dr.1. *See* Fokker aircraft
Drag, 24, 26, 142, 146, 196
Dresden, 134
Dunkirk, 113–114
D.Va. *See* Albatros aircraft
D VII. *See* Fokker aircraft

E

Eagle. *See* McDonnell Douglas aircraft: F-15 Eagle
Earhart, Amelia, 69, 80–82, 86–87; trophies, 72–73
Eastern Airlines, 46, 100
Edison, Thomas, 173
Eilmer, 16
Eindecker. *See* Fokker aircraft
Electra. *See* Lockheed aircraft
Elevators, 27, 30–31, 36
Ellsworth, Lincoln, 71
Endurance records, 84
English Channel crossings: first balloon, 18; first airplane, 39; first flown by woman, 64–65
Enola Gay, 138–139

F

F4F Wildcat. *See* Grumman aircraft
F-4 Phantom II. *See* McDonnell Douglas aircraft
F4U Corsair. *See* Vought F4U Corsair
F6F Hellcat. *See* Grumman aircraft
F-15. *See* McDonnell Douglas aircraft
F-16. *See* General Dynamics F-16
F-22. *See* Lockheed aircraft
F-80 Shooting Star. *See* Lockheed aircraft
F-86 Sabre. *See* North American aircraft
F-102 Delta Dagger. *See* Convair aircraft
F-106 Delta Dart. *See* Convair aircraft
F-117A Nighthawk. *See* Skunk Works
Fairey Delta 2, 145, 150
Fairgrave, Phoebe, 66
Farman, Henri, 38
Farman aircraft, 42
Fat Alberts, 201
Fat Man, 139
FB.5 Gun Bus. *See* Vickers aircraft
F.E.2b. *See* de Havilland aircraft
Fighter planes: development of, 48, 186–195; jets, 142–149; supersonic, 149–154; in World War I, 49–56; in World War II, 109–111, 113, 122
Fly-by-wire avionics, 170, 189
Flying boats, 68. *See also* Seaplanes
Flying circus, 51
Flying Tigers, 125
Flying wing, 193
Focke-Achgelis aircraft: FA-61, 175; FA-223, 175; FA-330, 175
Focke, Heinrich, 175
Focke-Wulf 190, 132
Fokker, Anthony, 38–39, 48, 71, 100
Fokker aircraft: C-2, 82; Dr.1, 41, 47, 51, 52; D VII, 56; Eindecker, 48–50; E.III, 48–49; F.VII-3, 71; Trimotor, 101; T-2, 71, 78
Folgore. *See* Macchi aircraft: MC-202 Folgore
Fonck, René, 47, 78
Ford, Henry, 97
Ford Tri-motor, 71, 101
Formation flying, 58, 112
France: and early airplanes, 36–37; in World War I, 42, 50; in World War II, 109
Franklin, Benjamin, 18
French Aero Club, 37
French, John, 43
Fuchida, Mitsuo, 120
F.VII-3. *See* Fokker aircraft

G

Galaxy. *See* Lockheed aircraft: C5A Galaxy
Garros, Roland, 48
Garuda, 17
General Dynamics F-16, 189
Germany: dirigibles in, 45, 49; jet development in, 142–143; in World War I, 42, 48–50; in World War II, 108, 112–115, 119–120
G forces, 148
Glamorous Glennis, 141, 151
Glenn Martin Company, 35
Gliders, 19-21
Gloster Meteor, 144
Göring, Hermann, 55, 113–116, 117
Goodrich Company, B. F., 85

Gordon Bennett Trophy, 34
Gossamer Albatross, 185
Gossamer Condor, 185
Gossamer Penguin, 185
Gotha G V, 58–59
Great Britain: early aviation in, 19; in World War I, 42, 50; in World War II, 112–120
Grumman aircraft: F4F Wildcat, 124–125; F6F Hellcat, 137; Gulfstream I, 165; X-29, 194
Guest, Amy, 80
Gulfstream I. *See* Grumman aircraft

H
I

H-1. *See* Hughes aircraft
Hall, James N., 49
Halsey, William, 121
Hamilton, Charles, 64
Handley Page 0/100, 58
Hands-on throttle and stick (HOTAS), 200
Harding, Warren G., 91, 92
Harrier. *See* British Aerospace Harrier
Hawaii, first flight to, 82
Hawker Hurricane, 109, 113, 116, 118
Hawk 75. *See* Curtiss aircraft
He 178. *See* Heinkel aircraft
Head-up Display (HUD), 198
Hegenberger, Albert, 82
Heinkel aircraft, 119; He 178, 142
Helicopters, 17, 173–177
Hellcat. *See* Grumman aircraft: F6F Hellcat
Henson, William S., 19
Herring, Augustus, 34
HH-3F. *See* Sikorsky aircraft
High-altitude flight, 85, 129, 137
Hiller, Stanley, 176
Hindenburg, 105
Hiroshima, 138–139
Hispano-Suiza engine, 54
Hitler, Adolph, 115, 119, 120, 142
HK-1 Hercules. *See* Hughes aircraft: *Spruce Goose*
Homebuilts, 182
Hoover, Herbert, 99
Houston, Lady Lucy, 67
HueyCobra. *See* Bell aircraft
Hughes aircraft, 176; 500D, 177; H-1, 70; *Spruce Goose,* 181
Hurricane. *See* Hawker Hurricane
Hustler. *See* Convair aircraft
Hydroaeroplanes. *See* Seaplanes
Hypersonic flight, 196–197
Icarus, 16
Immelmann, Max, 46, 50, 51
Imperial Airways, 95
Instruments, navigational, 94, 98. *See also* Avionics

J
K

Japan: bombing of, 135, 138–139; in World War II, 120–125, 135, 138–139
Jenny. *See* Curtiss aircraft: JN-4D (Jenny)
JetRanger. *See* Bell aircraft
Jets: development, 140–144; in Korean Conflict, 145–148; passenger, 156–172; supersonic, 149–154
Johnstone, Ralph, 65
Jones, R. T., 195
JU 87 Stuka. *See* Junkers aircraft
JU 88. *See* Junkers aircraft
Junkers aircraft: Ju 87 Stuka, 108, 114, 119; Ju 88, 114
Kaiser, Henry, 181
Kaman, Charles, 176
Kamikazes, 121, 138
Kate torpedo bomber, 120
Kelly, O. G., 71
King's Cup, 66
Kitty Hawk, North Carolina, 26, 28–29, 30
KLM, 95
Knight, Jack, 91–92
Korean Conflict, 145–148, 176–177

L

L-1011 TriStar. *See* Lockheed aircraft
Lafayette Escadrille, 49
Laminar flow control, 196
Lancaster. *See* Avro aircraft
Landing gear, retractable, 128
Langley, Samuel P., 23, 25
Law, Ruth, 64
Learjet, 164–165
Lear, William, 165
LeMay, Curtis, 135
Leonardo da Vinci, 17, 173
Liberty engine, 60, 83, 180
Lift, 19, 24, 30
Lighter-than-air flight. *See* Balloons; Dirigibles
Lightning. *See* Lockheed aircraft
Lilienthal, Otto, 20, 25, 26
Lincoln Beachey Fliers, 64
Lindbergh, Charles, 66, 73–79, 93, 137
Little Boy, 138–139
Lockheed aircraft: C-5A Galaxy, 156; C-130, 155–156; C-141A Staflifter, 156; Constellation, 157; Electra, 86; F-22, 204; F-80 Shooting Star, 144; L-1011 TriStar, 168; P-38 Lightning, 135, 136–137; Vega, 81, 82, 85; XP-80, 144
Locklear, Ormer, 66
L'Oiseau Blanc, 78
Louis XVI, 18
Lufbery, Raoul, 49
Lufthansa, 95
Luftwaffe, 108, 113–120

M

M-2. *See* Douglas aircraft
M-130. *See* Martin M-130
MacArthur, Douglas, 122
Macchi aircraft: C.202 Folgore, 132
MacCready, Paul, 182, 185
Mach, Ernst, 152
Machine guns, 48; Lewis, 50; Vickers, 54
Macready, J. A., 71
Magic windows, 200
Maitland, Lester, 82
Manly, Charles, 23
Mannock, Edward, 47
Marie Antoinette, 18
Martin M-130, 68, 98
Martin, Glenn, 35
McDonnell Douglas aircraft: DC-8, 156, 163–164; F-4 Phantom II, 149, 152; F-15 Eagle, 189
Me 109. *See* Messerschmitt aircraft
Me 110. *See* Messerschmitt aircraft
Me 163. *See* Messerschmitt aircraft
Me 262. *See* Messerschmitt aircraft
Memphis Belle, 134
Mercedes engine, 55
Mercury (god), 16
Merlin engine, 109–111, 118, 129
Messerschmitt aircraft, 70: Me 109, 108, 110–111, 113, 115, 117,118, 119; Me 110, 108, 119; Me 163, 116; Me 262, 142–144
Messerschmitt, Willy, 110
Meteor. *See* Gloster Meteor
MiG aircraft: MiG-15, 144, 146–148; MiG-21, 152
Mikoyan and Gurevich (MiG), 147
Mitchell bomber. *See* North American aircraft: B-25 Mitchell bomber
Mitchell, William, 61, 83
Mitsubishi Zero, 120, 122, 124
Monomail. *See* Boeing aircraft
Montgolfier, Etienne and Joseph, 18

Morane-Saulnier aircraft, 49; MS 406, 109
Morrow, Anne, 79
Moseley, Corliss, 67
MS 406. *See* Morane-Saulnier aircraft
Multi-axis thrust vectoring (MATV), 188. *See also* Thrust vectoring
Mustang. *See* North American aircraft: P-51 Mustang
Myths of flight, 16

N O

Nagasaki, 139
Nagumo, Chuichi, 124–125
National Advisory Committee for Aeronautics (NACA), 146
National Aeronautics and Space Administration (NASA), 197
National Air Pilots Association, 94
National Air Races, 66, 67
National Geographic Society, 82
Naval Fighter Weapon's School (Top Gun), 149
Navy-Curtiss (NC) flying boats. *See* Curtiss aircraft: NC
Newcomb, Simon, 22
Nieuport aircraft: Nieuport 11, 50; Nieuport 28, 56
Nighthawk. *See* Skunk Works F-117A Nighthawk
Nixon, Richard, 172
Noonan, Fred, 86
North American aircraft: B-25 Mitchell bombers, 123; F-86 Sabre, 146–148; P-51 Mustang, 110–111, 129, 145
Northrop Alpha, 101
Northrop, John, 191
Northwest Airlines, 100
Nungesser, Charles, 47, 78
Observer Corps, 116–117
Ohain, Hans von, 142
Optical fibers, 200
Ornithopters, 16–17, 178
Orteig Prize, 72, 78

P Q

P-40E Warhawk. *See* Curtiss aircraft
P-47 Thunderbolt. *See* Republic P-47 Thunderbolt
P-51 Mustang. *See* North American aircraft
P-59 Airacomet. *See* Bell aircraft
Pacific flights, 82
Pan American Airways, 98, 101, 163
Parachutes, 94
Passenger airlines: European, 94–95; jets, 158–172; legislation, 97–98; planes, 101–104, 206–207; United States, 96–97, 99–100
Patton, George, 134
Pearl Harbor, 120, 122
Pegasus, 16
Perseus, 16
Pershing, John, 47
Persian Gulf War, 156, 187, 189
Pfalz D.III, 52
Philippines, 122
Piasecki, Frank, 176
Pilâtre de Rozier, J. F., 18
Pilots: flying conditions for, 60, 92–93, 129; standards for, 98; in World War II, 118
Piper Cub, 104
Pitcairn Mailwing, 90
Pitch, 36
Polar flights, 71
Post, Wiley, 63, 85
Pratt & Whitney engines: J-57, 158; Wasp, 101
Pressurized aircraft, 136, 158–159
Prince, Norman, 49
Propellers, 27
Propfans, 197
Propulsion, 19, 27, 143, 196–197
Pulitzer Race, 67
Question Mark, 84
Quimby, Harriet, 64–65

R2C-1. *See* Curtiss aircraft
R3C-2. *See* Curtiss aircraft
R-6. *See* Curtiss aircraft
Races, air, 66–70
Radar, 114, 117, 132; and jets, 148; and stealth aircraft, 191–193
Radome, 148
Recreational flying, 182–183
"Red Baron." *See* Richthofen, Manfred von
Refueling, midair, 84
Reitsch, Hanna, 116
Remotely piloted vehicles (RPVs), 205
Republic P-47 Thunderbolt, 126, 129, 143
Richthofen, Manfred von, 41, 46–47, 55
Rickenbacker, Edward Vernon, 46–47
Rockne, Knute, 101
Rodgers, Calbraith, 65
Roll, 36–37
Rolls-Royce engines, 146; Avon, 152, 158; Merlin, 109–111, 118, 129; R (racing), 67
Roosevelt, Franklin Delano, 98, 100, 128
Royal Flying Corps (RFC), 50
Rudders, 27, 31
Rutan, Burt, 182, 184, 194
Rutan, Dick, 182, 184–185
Ryan aircraft, 73–74. *See also Spirit of St. Louis*

S

707. *See* Boeing aircraft
727. *See* Boeing aircraft
747. *See* Boeing aircraft
777. *See* Boeing aircraft
Sabre. *See* North American aircraft: F-86 Sabre
Saetta. *See* Macchi aircraft: MC-100 Saetta
Santos-Dumont, Alberto, 37
Saulnier, Raymond, 48
Schneider, Jacques, 67
Schneider Trophy, 66–67, 68, 70
Scott, Blanche, 64
S.E.5a, 51, 52
Sea Knight. *See* Boeing aircraft: CH-46 Sea Knight
Seaplanes, 67, 68
Seversky aircraft, 69
Shooting Star. *See* Lockheed aircraft: F-80 Shooting Star
Sikorsky aircraft: HH-3F, 177; R series, 175; S-55, 176; S-61, 175
Sikorsky, Igor, 39, 174–177
Skunk Works, 187; F-117A Nighthawk, 187, 192–193; SR-71 Blackbird, 187; U-2, 187
Smith, Lowell, 84
Smithsonian Institution, 23
Societé Pour l'Aviation et ses Derivés (SPAD), 58
Solar Challenger, 185
Sopwith aircraft: Camel, 52, 56; Snipe, 56; Tabloid, 57
Sound barrier, 141, 153
Spaceflight, 199
SPAD VII, 50, 54; SPAD XIII, 52, 54–55
Speed records, 67–70, 153, 199
Spins, recovering from, 64
Spirit of St. Louis, 74–79
Spitfire. *See* Supermarine aircraft, Spitfire
Spruce Goose, 181
SR-71 Blackbird. *See* Skunk Works
Stability, in flight, 17
Staggerwing. *See* Beech aircraft
Starlifter. *See* Lockheed aircraft: C-141A Starlifter
Stealth aircraft, 190–191
Stinson, Eddie, 64
Stinson, Katherine, 64
Stringfellow, John, 19
Stuka. *See* Junkers aircraft: Ju 87 Stuka

Stultz, Bill, 80
Stunt fliers, 64–66
Supermaneuverability, 201
Supermarine aircraft: Spitfire, 68, 112–113, 116, 118; S.6B, 67, 68
Supersonic transport (SST), 157–172

247. *See* Boeing aircraft
314. *See* Boeing aircraft
T-2. *See* Fokker aircraft
Tabloid. *See* Sopwith aircraft
Taft, William Howard, 32
Taube aircraft, 42, 49, 57
Taylor, Charles, 32
TBD Devastator. *See* Douglas aircraft
Thaden, Louise, 69
Thaw, William, 49
Thompson Trophy, 67, 72
Thrust-vectoring, 186, 188
Thunderbolt. *See* Republic P-47 Thunderbolt
Tin Goose, 101
Trainer planes, 35, 60
Transatlantic flights: Earhart crossing, 82; first airplane crossing, 68; first nonstop crossing, 72; Lindbergh crossing, 73–79
Transcontinental Air Transport (TAT), 81, 99
Transcontinental and Western Air (TWA), 99–100
Transcontinental (North America) flights, 65, 71
Travel Air Manufacturing Company, 99; Travel Air, 99
Trimotor. *See* Fokker aircraft
Trippe, Juan, 98
Truman, Harry, 138–139
Tupolev Tu, 144, 172
Type 75. *See* Curtiss aircraft

U
V

U-2. *See* Skunk Works
Udet, Ernst, 55
Ultralights, 182–185
United Aircraft and Transport Corporation, 99–100
U.S. Air Force Air Medal, 122
U.S. Post Office, 90–92, 97
U.S. Women's Airforce Service Pilots (WASPs), 133
V-22 Osprey. *See* Bell-Boeing V-22 Osprey
Val dive bomber, 120
Vega. *See* Lockheed aircraft
Vertical/Short Takeoff and Landing (V/STOL), 206
Vertical Takeoff and Landing (VTOL), 190
Vickers aircraft: FB.5 Gun Bus, 50; Vimy, 72
Vietnam, 149, 152, 154, 155, 177
Vimy. *See* Vickers aircraft
Vin Fiz, 65
Virtual cockpits, 200
Vishnu, 17
Voisin, Gabriel, 38
Voisin VIII, 42
Vought F4U Corsair, 137, 145
Voyager, 182, 184–185

W

Whitcomb, Richard, 195
Whitten-Brown, Arthur, 72
Whittle, Frank, 144
Wildcat. *See* Grumman aircraft: F4F Wildcat
Wilson, Woodrow, 90
Wind tunnels, 27
Wing design, 24; cantilever, 101, 144; current research in, 192–195; delta, 145, 152; early, 18, 21; and jets, 145; mission-adaptive, 194–195; oblique, 195; swept-back, 142, 145, 146, 153, 194, 195; swept-forward, 194; Wright brothers, 27; X-wings, 195
Wing-flapping devices. *See* Ornithopters
Wing stall, 24
Wing warping, 26
Wingtip-vortex turbines, 195
Women, in aviation, 64–66, 69, 80–82, 116; in World War II, 126, 133
Women's Air Derby, 69, 81
World Cruiser. *See* Douglas aircraft
World War I, 40–61; air tactics in, 51; balloons in, 44; bombing in, 57–61 mapmaking in, 43, 44; reconnaissance flights in, 43–45; dirigibles in, 44–45
World War II, 106–148; Battle of Britain, 115–120; Battle of France, 109, 112–114; Battle of Midway, 124–125; Battle of the Philippine Sea, 135; jets in, 142–144; Operation Sealion, 114–120; Pearl Harbor, 120, 122; War in the Pacific, 120–125, 135–139; women in, 126, 133
Wright aircraft: Wright 1903 Flyer, 28–31; Wright A biplane, 32; Wright Flyer II, 32
Wright, Orville and Wilbur, 22, 24-33; gliders, 25–27; first successful flight, 28–29; patent conflicts, 32, 36; exhibition team, 64; Whirlwind engine, 75, 90, 101

X
Y

X-1. *See* Bell aircraft
X-31, 186
XF-85 Goblin, 181
XP-80. *See* Lockheed aircraft
Yamamoto, Isoroku, 135
Yaw, 36–37
Yeager, Charles, 141, 151
Yeager, Frank, 93
Yeager, Jeana, 182, 184–185

Z

Zeppelin, Count Ferdinand von, 45
Zeppelins. *See* Dirigibles
Zeppelin-Staaken R VI, 59

Unless otherwise noted, all images have been provided by the National Air and Space Museum Library/Archives at the Smithsonian Institution. The image negative number is indicated.

Special thanks to The Boeing Company, Lockheed Corporation, McDonnell Douglas, Northrop Corporation, the United States Air Force Museum, and the Department of Defense Still Media Record Center.

DOD— Department of Defense; SI— Smithsonian Institution; NASM— National Air and Space Museum Library/Archives; USAF— United States Air Force Photo Collection, courtesy of SI, NASM; USAFM— United States Air Force Museum, Ohio

2–3 Seaver Center for Western History Research, Natural History #6978; **4–5** © Frank B. Mormillo; **6–7** DOD; **8–9** DOD; **11** Rockwell International; **15** © William B. Folsom; **16** Alinari/Art Resource, NY (t), Scala/Art Resource, NY (b); **17** Scala/Art Resource, NY (t), A-14756 (b); **18** 87-6624; **19** 85-18307; **20** A-49370E (t), A-39013 (b); **21** © Michael Freeman; **22** A-18840 (tl), 85-18303 (tr), Library of Congress (b); **23** 18827-B (t), A-18824 (b); **25** © Charles H. Phillips (tl), A-43268 (tr), 90-16502 (b); **26** A-6159 G (t), 86-3020 (b); **27** A-2708-G; **28–29** A-26767 (t), Wright Brothers National Museum (b); **30–31** 79-759; **32** 88-7998; **33** 42962-A; **34** © Charles H. Phillips; **35** A-5163-D; **36–37** A-48531-C (t), 87-10389 (bl), Biblioteca Nacional Do Rio De Janeiro, 94-S784 (br); **38** A-43517 C (t), © Frank B. Mormillo (b); **39** 78-14976 (t), 89-4479 (b); **41** © Frank B. Mormillo; **42** 94-3360; **43** 94-3366 (t), 94-3365 (b); **44** © Charles H. Phillips (t), SI Shell Co. Foundation, NASM Accession No. R 8 1965, 94-5782(b); **45** USAF 5207-AC; **46** 87-7706 (l), Dell Publishing Company, Inc., 94-6298 (r); **47** 94-7859 (l), 71-2903 (r); **48** Courtesy of Jim and Zona Appleby and THE FIRST WARPLANES™; **49** 89-19604; **50** © Frank B. Mormillo; **51** © Frank B. Mormillo; **52–53** USAFM; **54–55** A-43064 (t), 86-12094 (b); **56** USAFM; **57** Warren Bodie Collection, 81-14467; **58** 91-3474; **59** 91-17324 (t), USAFM (b); **60** © Frank B. Mormillo (t), © Charles H. Phillips (b); **61** Library of Congress; **63** 43061-B; **64** 84-18035; **65** 60-2081 (t), A-4485-C (b); **66** 76-2454; **67** 94-8398; **68** 60-2080; **69** 83-2144; **70** 60-4965; **71** A-44235 (t), A-4321 (b); **72** Photos by Mark Avino, 93-16197 (l), 93-16211 (c), 93-5513 (r); **73** Photos by Mark Avino, 93-5512 (l), 93-16200 (r); **74–75** 79-763; **76–77** A-12720-C; **78** © Charles H. Phillips; **79** A-4818-B; **80** 82-8683; **81** 73-4032 (t), 60-2082 (b); **82** © Charles H. Phillips; **83** A-48828; **84** USAF 11228-AC (t), USAF 11239-AC (b); **85** Lockheed; **86** 80-3190 (t), © Charles H. Phillips (b); **87** 86-147; **89** 94-6293; **90** 94-5582; **91** USAFM; **92** 93-1066; **93** United Airlines; **94** Musée Air France, 94-6297 (l), 91-19892 (r); **95** Lufthansa German Airlines, 94-6295 (l), 94-7953 (c), KLM, 94-6296 (r); **96** 78-2501; **97** 71-1-105; **98** 94-6313; **99** United Airlines (t, b); **100** © Frank B. Mormillo; **101** Boeing; **102–103** McDonnell Douglas; **104** 80-4969; **105** USAF 12293-AC (t), © Charles H. Phillips (b); **107** © Ross Chapple; **108** © Frank B. Mormillo; **109** 89-20606 (t), Imperial War Museum (Photo No. CH 3900), 94-7863 (b); **110–111** 80-2088 (t), 80-2090 (b); **112** 87-4680; **113** 80-2091; **114** 74-3031; **115** Imperial War Museum (Photo No. C 465), 85-7272; **116** 94-7860; **117** Imperial War Museum (Photo No. C 1869), 90-4395 (l), 71-3105 (r); **118** USAF 51201-AC; **119** 89-22018 (t), © Charles H. Phillips (b); **120** K-1694; **121** 85-7301; **122** © Charles H. Phillips; **123** 81-880 (t), USAFM (b); **124** 80-2093; **125** National Warplane Museum; **126** 94-7875 (t), K-1889 (b); **127** Lockheed; **128** USAFM; **129** © Frank B. Mormillo; **130–131** © ADAM Inc./Brian Silcox; **132** 80-2089; **133** USAF 160449-AC; **134** USAF 76576-AC (t), USAF I-23980-AC (b); **135** © Charles H. Phillips; **136** © Frank B. Mormillo; **137** © Frank B.

Mormillo; **138** USAF 164707-AC; **139** USAF 58450-AC (t), Boeing (b); **141** A-2013; **142** Photo by Dale Hrabak, 79-4620; **144** Photo by Dale Hrabak, 94-4518; **145** Airbus Industrie of North America (t), Rolls Royce Corporation (b); **146–147** USAFM (t), © Frank B. Mormillo (b); **148** Lockheed PI-8405; **149** McDonnell Douglas; **150** General Dynamics Convair; **151** 79-756; **152** © Robert Genat, ARMS Communications; **153** 94-7868; **154** Boeing; **155** 75-15167 (l), 94-7876 (r); **156** Lockheed; **157** Lockheed; **158** 90-9560; **159** DeHavilland; **160–161** Boeing; **162** Boeing; **163** United Airlines; **164** Learjet TC84-1051 (t), Boeing (b); **166** Chicago Department of Aviation; **167** Boeing; **168** Boeing; **169** British Airways; **170–171** British Airways; **172** Sovfoto; **173** 87-15490; **174** Sikorsky Aircraft; **175** Sikorsky Aircraft; **176** USAF 86336-AC; **177** courtesy United States Army Aviation Museum (t, b); **179** General Dynamics Convair; **180** A-18087; **181** SI Shell Co. Foundation, NASM Accession No. R 8 1965, 89-12606 (t), Hughes Aircraft (b); **182** AeroVironment; **183** EAA Aviation Center (t, b); **184** Microlon Inc./Voyager Inc; **185** AeroVironment; **186** Rockwell International; **187** Lockheed (t, b); **188** © William B. Folsom; **189** DOD (t, b); **190** DOD; **191** Canadair C-67611 (t), Photo by Dale Hrabak, 83-2946 (b); **192–193** Northrop (t), Lockheed (b); **194** DOD (t), Beechcraft (b); **196** Cessna; **197** NASA 86-HC-22; **198** DOD; **199** Photo by Dane A. Penland, 79-833; **200–201** U.S. Customs Service; **202–203** DOD; **204** Lockheed; **205** Israel Aircraft Industries Ltd. (t), Canadair (b); **206** DOD; **207** Boeing; **208** NASA S83-30237.**Glossary: ace** A-48746-A; **Ader's *Eole*** 85-7837; **Aeromarine** A-49337-A; **Airbus** 94-6308; **airmail** 80-2101; **Albatros** 94-6302; **autogyro** A-31079-A; **Avro** 94-6303; **balloons** 94-5783; **ball turret** USAF 21788-AC; **Beachey** 94-6319; **Beech 35** 94-6305; **Bell 47** 80-4271; **Bell, A. G.** A-31421; **Boeing** 75-12185; **Boelcke** 71-2915; **bombsight** USAF 1780-AC; **Bristol** 94-6299; **Byrd** 89-1208; **Caproni** 75-16322; **Cayley** 74-10605; **Chanute** 21147-B; **Charles** Musée de l'Air (Photo no. MA 887), 94-5785; **Chennault** 79-9003; **Cochran** 78-15317; **Cornu** 93-9668; **Corrigan** 94-6321; **Curtiss floatplane** 94-6304; **deHavilland** 94-7715; **Deutsche Lufthansa** Lufthansa German Airlines, 89-5475; **Doolittle** 77-14602; **Dornier** A-47017-A; **Douglas** A-2015; ***Enola Gay*** USAF 164707-AC; **Farman** 94-6318; **Fokker, A.** 75-6979; **Fonck** International Newsreel Photo, 85-12338; **Gordon Bennett Trophy** A-354; **Grahame-White** 94-6317; **Hughes** 81-16962; **Imperial Airways** 94-6309; **Johnson** 94-7714; **Johnstone** A-43519-C; **KLM** KLM, 94-6310; **Langley, S.P.** A-48458; **Lear** Learjet Inc.; **daVinci** A-3379-L; **Lewis gun** USAF 123149-AC; **Lilienthal** 73-2242; **Lindbergh** 78-12207; **Luke** USAF 147497-AC; **Mannock** Imperial War Museum (Photo No. Q60800), 85-12318; **Markham** 80-9027; **Martin** A-53897; **Messerschmitt, W.** Deutsche Aerospace Washington Inc., 94-6315; **Mitchell** A-44773E; **Montgolfier** A-180542; **Nakajima** 94-6306; **National Air Races** 94-6312; **Northrop Alpha** 80-2100; **Northrop, J.** 75-5442; **Nungesser** A-41913; **ornithopter** A-39059G; **Pan Am** 94-6307; **parachute** 94-6322; **photo reconnaissance** 80-14846; **Quimby** A-44401-A; **Richthofen** Imperial War Museum (Photo No. Q23917), 94-4459; **Ryan** 94-6300; **Santos-Dumont airship** A-212; **Sikorsky, I.** 89-2235; **Sopwith Triplane** 83-13355; **spat** USAF A-17962-AC; **Stinson** A-33444F; **Thaden** 75-1841; **Turner** 77-2695; **TWA** TWA, 94-6311; **WASP insignia** 94-6316; **Voisin** 94-6301; **wind tunnel** USAF 37772-AC; **Wright's Bicycle Shop** 86-9864; **Yeager, C.** A-3673; **Zeppelin poster** 86-10063; **Zeppelin, Count** A-4873.

Aircraft on Display

The National Air and Space Museum, Smithsonian Institution
6th St. & Independence Ave., SW,
Washington DC 20560
(202) 375-2000
Houses the milestones of aviation, including the Wright Flyer, *Spirit of St. Louis*, and Yeager's X-1 *Glamorous Glennis*.

The U.S. Air Force Museum
Wright-Patterson Air Force Base,
Dayton, OH 45433
(513) 255-3286
Military aircraft used by the air force in World War I, World War II, and beyond are on display in the oldest and largest military museum in the world.

The U.S. Army Aviation Museum
Fort Rucker, AL 36362
(205) 255-3169
Army helicopters used in World War II, the Korean War, the Vietnam War, and since are showcased.

The National Museum of Naval Aviation
U.S. Naval Air Station,
Pensacola, FL 32508
(904) 452-3606
Navy and marine aircraft on display cover the history of naval aviation.

The EAA Air Adventure Museum
EAA Aviation Center,
Oshkosh, WI 54903
(414) 426-4800
Focused on recreational aviation, the Experimental Aircraft Association displays light airplanes as well as restored World War I and World War II aircraft. Each July it hosts a massive fly-in.

The Pima Air and Space Museum
6000 E. Valencia Road,
Tucson, AZ 85706
(602) 574-9658
A desert climate allows Pima to preserve hundreds of historically significant commercial, recreational, and military aircraft outdoors.

The Air Museum Planes of Fame
7000 Merrill Avenue, Chino, CA 91710
(909) 597-3722
Restored World War II airplanes make this museum unique. Jets from the inception of jet technology through the Vietnam War are also displayed.

The San Diego Aerospace Museum
2001 Pan American Plaza, Balboa Park,
San Diego, CA 92101
(619) 234-8291
Over 60 air- and spacecraft show the history of aviation.

More About Aviation

Boyne, Walter J. *Silver Wings*. New York: Simon & Schuster, 1993.

Boyne, Walter J. *The Smithsonian Book of Flight*. Washington, D.C.: Smithsonian Books, 1987.

Boyne, Walter J., and Donald S. Lopez, eds. *Vertical Flight*. Washington, D.C.: Smithsonian Institution Press, 1984.

Ethell, Jeffrey L. *Frontiers of Flight*. Washington, D.C.: Smithsonian Books, 1992.

Park, Edwards. *Fighters: The World's Great Aces and Their Planes*. Charlottesville, Va.: Thomasson-Grant, 1990.

Serling, Robert J. *The Epic of Flight*. Alexandria, Va.: Time-Life Books, 1982.